Volume 4

THE PROGRESS OF SCIENCE

T0262792

THE PROGRESS OF SCIENCE

An Account of Recent Fundamental Researches
in Physics, Chemistry and Biology

J. G. CROWTHER

Routledge
Taylor & Francis Group

LONDON AND NEW YORK

First Published in 1934

This edition first published in 2014
by Routledge
2 Park Square, Milton Park, Abingdon, Oxfordshire OX14 4RN

and by Routledge
711 Third Avenue, New York, NY 10017

Routledge is an imprint of the Taylor and Francis Group, an informa business

First issued in paperback 2015

British Library Cataloguing in Publication Data
A catalogue record for this book is available from the British Library

ISBN 978-0-415-73519-3 (Set)
ISBN 978-1-138-01350-6 (hbk) (Volume 4)
ISBN 978-1-138-98973-3 (pbk) (Volume 4)

Publisher's Note
The publisher has gone to great lengths to ensure the quality of this book but points out that some imperfections from the original may be apparent.

Disclaimer
The publisher has made every effort to trace copyright holders and would welcome correspondence from those they have been unable to trace.

THE PROGRESS OF SCIENCE

AN ACCOUNT OF
RECENT FUNDAMENTAL RESEARCHES
IN PHYSICS, CHEMISTRY AND BIOLOGY

BY

J. G. CROWTHER

LONDON

KEGAN PAUL, TRENCH, TRUBNER & CO., LTD.

BROADWAY HOUSE, 68–74 CARTER LANE, E.C.

1934

1st Edition	.	.	.	.	.	.	. *March*	1934
2nd Edition	.	.	•	.	.	.	. *November*	1934

Printed in Great Britain by Butler & Tanner Ltd., Frome and London

CONTENTS

LIST OF PLATES

PREFACE
TO THE FIRST EDITION

THIS book is a further contribution towards the information of the general reader concerning the progress of science. A knowledge of science is now necessary in every person who wishes to take an intelligent interest in human affairs. Scientific discovery has placed hitherto unknown powers into the hands of persons who rarely understand these powers in principle or in spirit, and consequently bungle the use of them in practice. Every person nowadays who has executive influence in politics and industry cannot use his influence intelligently unless he has a knowledge of scientific method. Political and industrial leaders can no longer leave problems involving scientific understanding to technical experts ; they must try to form their own opinions. If they do not, they become victims of false theories in philosophy, economics, biology, and even in physics. Innumerable persons believe there are short cuts to knowledge, that instinct will solve problems of psychology better than rational research, that it is permissible to believe without investigation and proof that white peoples are better than black, that commercial experts can assess the desirability of schemes of electrification and other vast engineering projects without any knowledge of electrical engineering, etc.

This book is intended to help the general reader to increase his knowledge of science and to make him more impatient of leaders who cannot take a scientific view of human problems. When the scientific spirit is widely diffused among humanity scientifically-minded leaders will be called for. The problems of a civilization based on applied science cannot be solved by leaders and participators without scientific understanding.

Mr. W. T. Astbury, Professor P. M. S. Blackett, Professor Lancelot Hogben, Professor N. F. Mott and Dr. Joseph Needham have kindly read parts of the manuscript. I am much indebted to them for reducing the number of errors. Professor Blackett, Dr. G. P. S. Occhialini, Dr. J. D. Cockcroft, Professor Niels Bohr, Dr. P. I. Dee, Dr. E. T. S. Walton, Dr. N. Feather, Dr. Honor Fell, Dr. R. G. Canti, Dr. P. Kunze, Professor E. O. Lawrence, Professor O. Mangold, Dr. Obreimov, Professor E. Regener, Dr. D. Skobelzyn, the Royal Society of London, and the proprietors of the *Zeitschrift für Physik* have kindly given me permission to reproduce illustrations.

J. G. C.

NOTE ON THE SECOND EDITION

I AM indebted to the Editor of the *Nineteenth Century* for permission to use the material of an article in the preparation of an additional chapter on *Artificial Radioactivity*.

J. G. C.

November 1934.

THE PROGRESS OF SCIENCE

CHAPTER I

THE CAVENDISH LABORATORY

I

THE series of discoveries made in the Cavendish Laboratory is the most remarkable contributed by one institution to contemporary science. No institution in history has had a more brilliant record. Its fame creates the illusion that it is old, yet it was opened in 1874, within the lifetime of many living scientists. It has had four directors only; but what a marvellous group! Clerk Maxwell was the first. Many would say he was the greatest scientific genius of the nineteenth century, he was less pedestrian than Helmholtz and Kelvin, and conceivably could have been surpassed only by Darwin and Faraday. Perhaps the first place is to be shared by Darwin and Faraday; nevertheless, Maxwell is close to them. As the founder of a tradition he was their superior. He could inspire others besides accomplishing his own brilliant work. Leadership was not among Faraday's stupendous talents, for he never had a scientific assistant, and was unable to perceive undeveloped talent. On behalf of the Royal Institution he was charged to discover talent which might be employed in the laboratory, but he failed to discover any suitable man after years of conscientious watching. Darwin also was an extreme individualist. His delicate

B

health, natural inclinations and fortune impelled him to work in quiet isolation. He did not communicate his intellectual habits to a school, he lived apart as the ancient wise men lived, and eminent disciples delivered without sufficient direction their own interpretations of the master's work. Darwinism came to have a content different from the ideas and method of Darwin. Maxwell's success in establishing the intellectual tradition of the Cavendish Laboratory was of a sort beyond the powers of Darwin and Faraday, and is an important part of his achievement. His talent for leadership did not infringe on his personal work. Some important figures in science have had this power of leadership without much personal talent: the distinguished physiologist Michael Foster created a wonderfully fruitful tradition in his science without making any discoveries of great value himself. The personal achievements and personality of Maxwell have caused his contribution as the founder of a tradition to be overlooked. His humour and his lack of pomposity prevented him from becoming a great Victorian, nor did he live long enough. Everyone knew he was intellectually brilliant, but was he sound and solid? He died before the period had decided. Maxwell's application to the problems of the new laboratory and organizing its routine has not received adequate praise. His health was weak and he was young enough to have been absorbed in his theoretical researches. Few would have blamed him if he had regarded the laboratory as an appendage of the study from whence profound theoretical researches came, where a tired brain might be soothed by manual distraction. Maxwell was a chief moulder of physical thought. He broadened the concept of the ether to describe all the known phenomena of magnetism and electricity and to suggest the existence of unknown phenomena, such as the electro-magnetic waves now applied in radio-communication. The success of his development of Faraday's idea of electro-magnetism as a disturbance in the ether increased the curiosity concerning this ether. How was it affected by the proximity of material bodies and was it at rest relative to them? The Michelson-Morley experiment was devised to answer these questions and the answer given required the invention of the theory of relativity to make it intelligible. Maxwell's explanations of the properties of gases and heat by applying the laws of statistics to assemblages of atoms led through Boltzmann to Planck's invention of the quantum theory of action. Maxwell was a direct ancestor of the two chief theoretical achievements of recent culture.

In the new Cavendish Laboratory he started researches on the exact determination of various constants of nature such as the unit of resistance to electricity. These researches extended until a special institution was created to continue them. Maxwell was one of the ancestors of the National Physical Laboratory. He died five years after the Cavendish Laboratory had been opened and within those years he had given it a tradition. He was succeeded by Lord Rayleigh.

The reputation of Rayleigh increases with time, as Maxwell's. He was extremely good both as theoretician and experimenter, and always attempted the most difficult problems. During his short tenure of the professorship, from 1879 to 1884, he continued the work on units and introduced the team method of research. He believed the laboratory as a whole should have some special line of investigation to which everyone could give some thought. He arranged afternoon tea intervals for rest and collective discussion which have had such a formative influence on many young physicists. He believed in using some routine investigation as a stand-by during periods of intellectual sterility. Many undistinguished but useful types of routine research serve to occupy the stale scientist. If he feels he is progressing, however modestly, he is saved from the attacks of psychological depression that attack the baulked intellectual. The accumulation of routine results may also cause the discovery of facts of unexpected importance. Rayleigh made the weighing of gases one of his 'stand-by's'. He believed that extremely accurate measures of the masses of atoms might help to discover evidence for the existence of a fundamental material of which all atoms were packets. His belief in some form of Prout's hypothesis has been justified and his routine weighing of gases led to the brilliant and unexpected discovery of the inert gases. This was not Rayleigh's only successful 'stand-by'. His continued studies of the properties of the surfaces of liquids had unforeseen value in recent investigations of monomolecular layers of substances, layers one molecule thick. The properties of monomolecular layers are important in the chemical phenomenon of catalysis in which a substance stimulates the reaction between others.

Catalysis is much used in chemical industry and may soon become the most important of chemical manufacturing methods. The gas-filled electric lamp was invented with the assistance of a knowledge of monomolecular layers. While Rayleigh was Cavendish professor successful courses of instruction in experimental physics were organized. They were given by Glazebrook and Shaw, who introduced some of the methods Pickering used in the Massachusetts Institute of Technology. Rayleigh resigned the chair in 1884. He could afford a private laboratory and was not disposed to perform the routine of an academic appointment unnecessarily. He was succeeded by J. J. Thomson, then twenty-eight years old. Thomson has modestly described how this appointment surprised him and the rest of the University. He relates how a well-known college tutor commented that things had come to a pretty pass in the University when mere boys were made professors.

Thomson had already shown exceptional theoretical power, as he had published many mathematical researches and had made one great discovery. In 1881 he proved that a moving charge of electricity behaves as if it had mass. This was the first considerable contribution towards the establishment of the electrical theory of matter, that matter is made of electricity. In the early part of his career his theoretical much exceeded his experimental achievements. Thomson's thought was never inspired by manual intuition, experiment was always the assistant of theory. There was no brilliant experimental opportunism in his long researches on the nature of the conduction of electricity through gases. They were conducted rationalistically, as a general may employ detailed methods in working out all possibilities and preparing for them, rather than detecting by intuition the probability among many possibilities. Thomson was not inspired by his instruments to discover entirely unsuspected phenomena, he had no pure experimental genius. His powerful mind conceived explanations which might be tested by others as capably as by himself. This shows his conceptions were in his conscious mind. Many great experimenters work with semi-conscious ideas which seem to become clearer and more conscious as they make their experiments.

Thomson's power of conceiving a large number of conscious ideas which could be tested was a factor in his unprecedented success as a laboratory director. He could find something definite for everyone to try. The capaciousness of his mind enabled him to co-ordinate many of these activities. Without being a supreme experimenter himself Thomson could conceive powerful experiments and direct the performance of them by others. He had chosen the problem of the conduction of electricity through gases as his main experimental research. He set his students to investigate various aspects of this complex problem and the infinitude of the complexities were an inexhaustible source of material for the active minds of the young men who came to his laboratory. The development of the team method of research in experimental physics was largely due to Thomson's combination of qualities. Kelvin's immense prestige did not enable him to create a tradition of team research at Glasgow. Kelvin was regarded as a super-man to be tended with religious awe. He was not much interested in teaching and used the help of his colleagues to study the problems in which he was interested. He was a hero whereas Thomson was a general.

The relative importance of Thomson's own discoveries and of his influence as the director of a school of research is difficult to assess. Both had an enormous influence on physics. The number of distinguished physicists whose investigatory powers were developed in Thomson's laboratory is extraordinary. Rutherford, Bragg, C. T. R. Wilson, Langevin, Callendar, are among the fifty professors of physics who at some date had studied under Thomson.

The discovering of the electron was his most famous achievement. Others independently discovered it about the same time but Thomson and his colleagues were the most successful explorers of its meaning and implications. His colleague Richardson systematically studied the emission of electrons from hot bodies, the phenomenon upon which the existence of the thermionic or radio-valve depends.

Thomson directed the Cavendish Laboratory for thirty-five years, from 1884 until 1919. Its reputation had become magnificent beyond the hope of sustainment but mortal expectation has been disproved by the last thirteen years of its history. Thomson was succeeded by Rutherford. The new professor had previously accomplished three major achievements, he had elucidated the phenomena of radioactivity, he had deduced the true structure of atoms with the assistance of these phenomena, and with their assistance he had accomplished the first artificial disintegration of an atom. He had already founded recent atomic physics; could more be expected of any man?

Rutherford organized a refined attack on many aspects of atomic structure. The outlines he had already drawn were to be exactly traced. As the atomic details continued to be studied the students increased their technical skill. The visual, the electrical, the photographic, the magnetic technique for investigating atomic phenomena were improved until the Cavendish Laboratory possessed an unequalled stock of skill in the experimental study of atomic problems. Kapitza's new methods of producing intense magnetic fields by short-circuiting a dynamo and sending an enormous momentary current through a coil had proved successful. A new technique had been invented which promised a stream of investigations occupying many years of research.

The fruits of the years of patient research were particularly notable in 1932. Dr. J. Chadwick discovered the neutron, Dr. J. D. Cockcroft and Dr. E. T. S. Walton succeeded for the first time in disintegrating atoms by machinery and without the assistance of radioactive substances, and Mr. P. M. S. Blackett devised the first camera for automatically photographing the tracks of cosmic rays, i.e., making cosmic rays operate a camera which would photograph their own tracks.

The smallest particles in nature have proved difficult to find.   Years of research by many workers were needed for the discovery of the elementary particle of negative electricity, the electron.   An elementary particle of positive electricity was not clearly recognized for more than a decade after the discovery of the electron in 1897.   Now that we know the hydrogen atom consists of but two parts, a nucleus consisting of a proton or unit of positive electricity, and a distant electron which revolves round the nucleus, the immense labours necessary to discover a proton and an electron seem extraordinary.   Research so often ends in this fashion.   The objects of a search conducted through the whole realm of nature by the ablest minds of several generations are found to be neatly packed together as one of the commonest things of experience.   Electrons and protons are produced with ease and in any quantity by treating hydrogen violently.   An electric spark in hydrogen gas will cause large numbers of hydrogen atoms to split into electrons and protons.   These elements of electricity are so easy to obtain—when you know how. The history of science offers one impressive lesson : the difficulty of discovering even the simplest unknown things.   If simple phenomena prove so difficult to understand, what evidence is to be given to confident assertions concerning complex phenomena?   By 1911 the proton had been clearly recognized.   The hydrogen atom is a loose combination of an electron and a proton. As the electric charges of an electron and a proton are equal and opposite, the electric charge of a normal hydrogen atom is neutral. The neutral atom is vastly larger than either a proton or an electron, which are much of a size. The electron is 4.10^{-13} cm. and the proton is 2.10^{-16} cm. in diameter, while the hydrogen atom is 2.10^{-11} cm. in diameter. Protons and electrons are thousands of times smaller than atoms, so the proton and electron in the hydrogen atom are relatively far apart. Physicists speculated

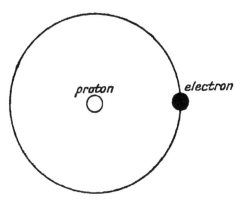

FIG. 1.   The hydrogen atom.

on the possibility of the existence of a particle consisting of an electron and proton in close connection, contrasted with the distant connection in the hydrogen atom. Rutherford was the first to give serious consideration to this possibility. Bragg had discussed in 1904 the possibility of neutral doublets consisting of a negative and a positive charge of electricity, in order to explain the phenomena of radiation, but he had been interested in inventing a theoretical explanation which would fit the phenomena without other evidence of its actuality. Rutherford's ideas were expressed in 1920, when the conceptions of the electron and proton were clearer and a neutral doublet as a combination of a proton and an electron had become a more substantial possibility. Glasson and Roberts made a search for this neutral close combination of a proton and an electron, or neutron, though without success. In 1920 Rutherford delivered the Royal Society's Bakerian Lecture, choosing for his subject the Nuclear Constitution of Atoms. He had achieved the first artificial disintegration of an atom in the previous year and the moment was suitable for a review of the state of knowledge of atomic structure. After describing the results of the previous decades of research Rutherford made some brilliant speculations concerning unknown sorts of atoms. Under the inspiration of his recent triumphs he said:

"The idea of the possible existence of an atom of mass 1 which has zero nucleus charge (is involved). Such an atomic structure seems by no means impossible. On present views the neutral hydrogen atom is regarded as a nucleus of unit charge with an electron attached at a distance, and the spectrum of hydrogen is ascribed to the movements of this distant electron. Under some conditions, however, it may be possible for an electron to combine much more closely with the hydrogen nucleus, and in consequence it should be able to move freely through matter. Its presence would probably be difficult to detect by the spectroscope, and it may be impossible to contain it in a sealed vessel. On the other hand, it should enter readily the structure of atoms, and may either unite with the nucleus or be disintegrated by its intense field, resulting possibly in the escape of a charged hydrogen atom or an electron or both."

This is one of the most remarkable predictions in the history of science. Recent discoveries have shown every sentence is significant. Rutherford continued:

"If the existence of such atoms be possible, it is to be expected that they may be produced, but probably only in very small numbers, in the electric discharge through hydrogen, where both electrons and hydrogen nuclei are present in considerable numbers. It is the intention of the writer to make experiments to test whether any indication of the production of such atoms can be obtained under these conditions. The existence of such nuclei may not be confined to mass 1, but may be possible for masses 2, 3, 4 or more, depending on the possibility of combination between the doublets. The existence of such atoms seems almost necessary to explain the building up of the nuclei of heavy elements, for, unless we suppose the production of charged particles of very high velocities, it is difficult to see how any positively charged particle can reach the nucleus of a heavy atom against its intense repulsive field."

The attempts to produce neutrons by electrical discharges in hydrogen were not successful. The subsequent successful discovery of neutrons of mass 1 enhances the interest of the suggestion that neutrons of higher mass may also exist. Evidence for their existence has already been claimed. Heavy neutrons might be of great value, as an increase in the mass of a bombarding particle more than proportionately increases its effectiveness. Refinement of radioactive investigatory technique combined with a fortunate acquisition of resources led to the discovery of the neutron. Dr. N. Feather, Dr. J. Chadwick's colleague, happened to have worked in America. He was given a number of old radon tubes by Dr. G. F. Burnam and Dr. F. West of the Kelly Hospital, Baltimore. These provided a more powerful source of polonium than that available in Cambridge. Polonium is specially useful because it emits pure beams of helium nuclei. Bombardment methods used by Rutherford and Chadwick in 1912 had been devised to study what happens when the particles ejected by radioactive substances strike the nuclei of other atoms. Rutherford had shown that the nuclei of atoms were small even if relatively massive. The nucleus of the atom of iron, for instance, was over fifty times as heavy as a proton, so it was made up, presumably, of over fifty protons and more than a score of electrons. This little multitude of objects within a compact nucleus would endow it with complexity. If such a nucleus were bombarded by particles from radium it might have pieces knocked out of it (Rutherford demonstrated in 1919 the possibility of knocking pieces out of the nuclei of aluminium atoms) or be caused to vibrate and emit special radiations. This was proved by J. A. Gray in 1911 and Chadwick in 1912. Chadwick was a young research student at Manchester University aged twenty-one years, when he showed that the nuclei of atoms bombarded by particles ejected from radioactive preparations might emit radiations of high frequency. He discovered the neutron twenty years later by experiments similar in principle to those of his early researches. The discovery of the possibility of excitation of atomic nuclei by electrons ejected from radioactive substances has been employed to investigate the structure of the nuclei. By striking the nucleus as if it were a bell and studying the notes of the emitted vibrations something of its structure may be deduced. As radioactive substances may emit two sorts of particles, electrons and nuclei of helium atoms, different sorts of effects may be obtained from nucleus bombardment according to which is used. The helium nuclei are nearly 8,000 times as massive as the electrons, so they may be more disturbing. In fact the first artificial disintegrations of atomic nuclei were done with these heavy particles and not the electrons. Both methods of bombardment have been studied carefully with increasing refinement as progress in the design of instruments has permitted greater accuracy. The German physicists Bothe and Becker have been engaged in these studies. In 1930 they discovered that the metal beryllium emitted peculiar radiations when bombarded with helium nuclei from a special radioactive preparation. Their radioactive source was strong and produced fast projectiles. These fast helium nuclei usually disintegrated the nuclei of light atoms if they hit them, knocking particles out of them. Bothe and Becker discovered beryllium did not appear to emit particles when bombarded, but a very penetrating radiation. They assumed it must be a radiation rather than a stream of particles because it could penetrate layers of lead inches thick. This seemed quite beyond the power of any known particle.

Bothe's beryllium rays were much more penetrating than any rays hitherto obtained from radioactive substances; their penetrating power was exceeded only by that of the cosmic rays. The students of cosmic rays were immediately interested because the beryllium rays seemed to be intermediate in power between radioactive radiations and cosmic rays. If they came within the band of unknown frequencies they might help to explain the problem of the cosmic rays whose nature is obscure. The beryllium rays became the object of keen study. Mme. Curie-Joliot, the daughter of Mme. Curie, and her husband M. Joliot investigated them at Paris and Dr. Chadwick and his colleagues at Cambridge. The Curie-Joliots made a striking discovery. They found the beryllium rays could knock particles out of paraffin wax and other substances containing hydrogen. The ejected particles proved to be protons moving at an extraordinarily high speed, from the slight degree in which they were deviated by a powerful magnetic field. This was very strange. If the beryllium rays were a wave-radiation their energy could be deduced from the speed of the ejected protons. The laws of the exchange of energy and momentum between wave-radiations and particles are known. Calculation showed that the beryllium rays, if they were waves, must have the extraordinary energy of 50,000,000 electron-volts. As the beryllium rays were themselves produced by the impact of helium nuclei on beryllium nuclei the energy of the bombarding particles could be calculated from the energy of the beryllium rays they produced. But the energy of the bombarding particles was already known, because they were obtained from polonium. This radioactive substance is convenient in these experiments because it emits helium nuclei only and no electrons or wave-radiations. The experimenter is not confused by the presence of other exciting rays. It happens that polonium was discovered by Mme. Curie-Joliot's mother, and named after her native country, Poland. The energies determined by direct measurement and by calculating backwards from the observed energy of the protons ejected from the wax did not agree. Curie-Joliot and Joliot deduced that the law of inter-action between waves and particles did not apply to the interaction between beryllium rays and protons. They believed they had discovered a new type of interaction between wave-radiation and matter to which the law of the conservation of energy did not apply. Meanwhile, Chadwick and Webster had discovered anomalies concerning the intensity of the beryllium rays in various directions. They were produced by helium nuclei and their intensity along the direction of the nuclei which had excited them was much greater than in the reverse direction. The difference was greater than could be explained if the beryllium rays were waves.

The behaviour of the beryllium rays was anomalous and apparently did not obey the law of the conservation of energy. Curie-Joliot and Joliot had invoked a new type of interaction between radiation and matter to explain the observations. Then Chadwick was inspired to consider whether the beryllium rays might be streams of neutrons. He had been Rutherford's close collaborator for many years, he had often discussed the existence of neutrons and searched for them, he was working with Rutherford when the remarkable speculations of 1920, which have been quoted, were published. This is not the first occasion upon which Continental physicists have assumed rays to be waves which later were proved by British physicists to be particles. The British mind has an aptitude for detecting the particulate aspect of natural phenomena. Perhaps it is due to a tradition of ball-games which trains the imagination in the behaviour of flying particles, so that when they are met in the laboratory they are swiftly recognized. The history of the cathode rays is a notable example. British investigators tended to assume they were particles, while Continental investigators assumed they were waves. Newton supposed light to consist of particles, Huyghens and the Continentals supposed it to be a wave-phenomenon. Faraday's researches on the passage of electricity through conducting liquids implied that unit quantities of electricity were associated with atoms, but the implication that electricity is particulate was drawn by the German physicist Helmholtz. Faraday had little belief in the atomic theory of matter, which had been chiefly invented by another Englishman, Dalton. The modern wave-theory of matter is the invention of de Broglie, a Frenchman, and the leading theoreticians of wave-mechanics are nearly all Continentals. The exception, Dirac, helps to prove the rule, as he is of French ancestry, though born in England.

When the helium nuclei strike the beryllium nuclei the former are probably incorporated with the latter. The Curie-Joliots suggested the two nuclei combined to form the nucleus of a carbon atom of mass 13. The ordinary carbon atom is of mass 12. According to them the reaction could be described:

beryllium nucleus + helium nucleus
   of mass 9   of mass 4
        = carbon nucleus + the wave-radiation
        of mass 13

The masses mentioned in this equation are not exactly units. The sum of the masses on the left-hand side would be slightly less than the mass of the carbon nucleus. The small difference would represent the transformation of mass into energy which would explain the origin of the wave-radiation. As Einstein showed, mass and energy are equivalent according to the modern theory of matter. Mass may be changed into energy in the form of waves. The light and radiative energy of the sun and stars is produced by the transformation of their material into wave-energy. Even a candle or a searchlight radiates by virtue of this transformation, though the quantities of mass so transformed are excessively minute and beyond direct measurement but not calculation. Atoms of matter are an extremely concentrated form of energy; the celestial haberdashers' convenient packets of the fundamental commodity of the universe. If the equation were correct the known slight difference of mass would imply that the wave-radiation was of an energy equivalent to 14×10^6 electron volts. But the rays produced, if they are waves, must be emitted as quanta of energy equivalent to 50×10^6 electron volts. The observed beryllium rays were three times as energetic as the loss in mass implied.

The Curie-Joliots interpreted this as a departure from the hitherto established laws of the interaction of radiation and matter.

Dr. Chadwick showed all of these contradictions could be resolved if the beryllium rays were supposed to be neutrons and not waves. The equation of interaction would be written thus :

beryllium nucleus + helium nucleus
     of mass 9   of mass 4
           = carbon nucleus + neutron of
                 of mass 12   mass 1

The production of an ordinary carbon nucleus is more probable. The slight transformation of mass

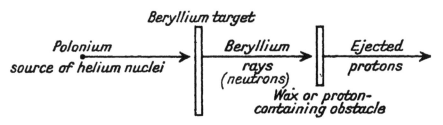

FIG. 2.   The production of neutrons.

will endow the neutron with approximately sufficient energy to strike protons out of wax with the speed observed by the Curie-Joliots. The conservation of energy and momentum is obeyed. The difference in the speed of protons ejected by the beryllium rays and of those travelling in the same direction as the original exciting polonium rays is also explained because the neutrons as particles would naturally knock forward other particles in their way.

The evidence for the existence of the neutron depends on a satisfactory explanation of the speeds of various particles after collisions.

Chadwick's theory adequately explains the observed speed of the proton and the probable interactions represented in Fig. 3. The success in interpreting the results not of one but of three consecutive collisions shows the degree of the improvement of experimental technique. These particles are less than one million-millionth of a centimetre in diameter and the helium nucleus, the neutron and the proton are moving at about 20,000 miles a second.

The beryllium rays were assumed to be waves because they pass through air without colliding with many atoms. Rays consisting of particles such as electrons, protons or helium nuclei collide with

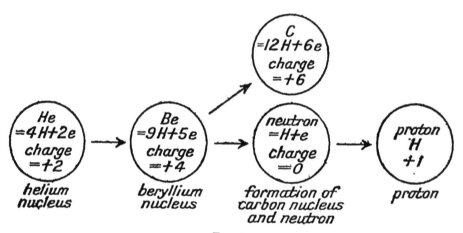

FIG. 3.

thousands of atoms as they pass through a few centimetres of air. All of these particles have positive or negative electric charges which repel or disturb the atoms they approach. They ionize the atoms of air, removing some of the outer electrons of the oxygen and nitrogen atoms. Neutrons pass through the atoms without removing the electrons unless they strike an oxygen or nitrogen nucleus directly. This is shown by the remarkable photographs obtained by Dr. N. Feather (Plate 1).

He found short fat tracks of struck atomic nuclei might be obtained without any visible track leading

PLATE 1

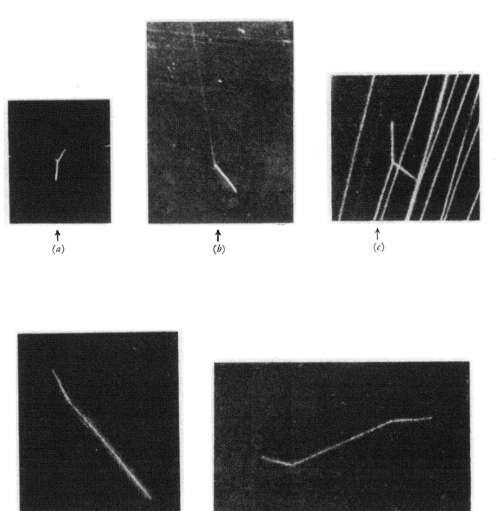

Photographs of neutron disintegrations by Dr. N. Feather.

(a) A neutron has struck a carbon atom and knocked it forward. The struck carbon atom has then bumped into a second carbon atom, making a forked track. The neutron, which passes through the gas according to the direction of the arrow, leaves no track.

(b) A neutron has struck and disintegrated a nitrogen atom. The neutron makes no track. The thin track is of a synthesized helium atom, and the thick track of a boron atom.

(c) Compare this forked track with those in (a) and (b). The impinging particle is in this case a helium atom, not a neutron, which has struck another helium atom. It makes a track both before and after impact.

(d) A neutron has disintegrated an oxygen atom. The track to the right is possibly of a helium atom, and the short track to the left is of a carbon atom.

(e) A neutron has disintegrated an oxygen atom, probably with the same result as in (d). The track to the right may, however, be of a heavy hydrogen, or diplogen, atom. Then the track to the left would be of a nitrogen atom. The atom that made the track to the right, whether helium or diplogen, has collided with another atom towards the end of its path.

(N. Feather and Royal Society.)

to their starting ends. The neutron had passed through the air without encountering any obstacle until it had struck the nucleus of an atom. It had not interacted with the far more numerous electrons in the outer parts of the atoms of oxygen and nitrogen. Why did the neutron miss the electrons and continue until it hit a nucleus? This could be explained roughly according to Rutherford's ideas of 1920. The neutron has no electric charge, so it may approach an electron very closely without disturbing it. If it were surrounded with the aura of an electric field it would brush electrons off atoms as protons or other charged particles do. It has not this aura and therefore has a much less effective size and correspondingly greater penetrating power. It goes through substances because it misses most of the obstacles. Yet, if this explanation were complete some collisions with electrons would be probable. Little definite evidence for such collisions has been obtained; less than expected. Neutrons from beryllium are capable of penetrating several feet of lead and upwards of one mile of atmospheric air. Protons and other comparable particles penetrate only a film of lead and a few centimetres of air. The extraordinary penetrating power of neutrons is due to the rarity of their interactions with electrons. They do not lose energy in penetrating matter by colliding with the multitudes of neighbouring electrons. While the rarity of interactions may be foreseen on general grounds, as Rutherford had predicted, the observed extreme infrequency was not explicable exactly. Within a few weeks of Chadwick's announcement of the probable existence of neutrons Niels Bohr had given a characteristically ingenious solution of the difficulty. It is based on the wave-theory of matter. According to the present wave-theory of matter particles consist of wave-packets. These waves are not of the same sort as light waves or radio waves, but consist rather of waves of probability of position. Particles behave as if their motion were controlled by waves of probability, which makes them appear as if they were made of some sort of waves. The wave-like characteristics of particles cause them to exhibit some of the phenomena of waves. For instance, waves of equal size but travelling in opposite directions may obliterate each other at some points when they meet, and reinforce each other at other points. Waves are reflected or stopped by obstacles as large or larger than themselves but obstacles smaller than themselves do not stop

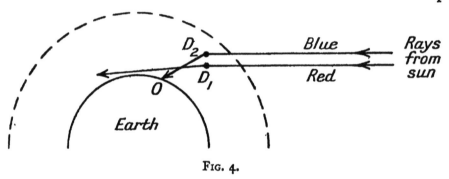

Fig. 4.

them. Water-waves wash over small stones in their path without being modified.

The red ray is diffracted less by the particle of dust at D_1 than the blue ray by the particle of dust at D_2, consequently the sky appears blue to the observer at O.

The blue rays starting towards the observer are diffracted sideways, so he sees only the less diffracted red rays.

This phenomenon explains the blueness of the sky and the redness of sunsets. The light from the sun illuminates the air overhead by many rays which pass through obliquely, as represented in Fig. 4. If the oblique rays are of long wave-length, or red light waves, they will not be deflected or diffracted much, because their waves will be larger than the particle of dust and molecules in their path through the air. If the oblique rays are of short wave-length, or blue light waves, they will be diffracted more, because the particles in their path will be more nearly of their size. Thus blue rays will be diffracted more than red rays by dust particles such as D_2, to reach an observer O, who will receive the impression that the sky is blue because the majority of the rays reaching his eye are blue. Similar explanations show how the sun appears red at sunset because the particles of dust diffract not the oblique but the direct rays

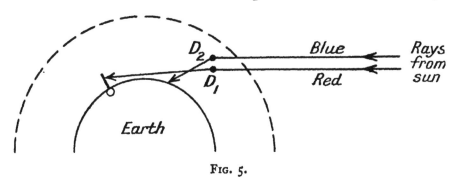

FIG. 5.

from the sun. The blue rays are unable to reach the observer because they are diffracted by particles of dust such as D_2 in Fig. 5, while the red washes over obstacles such as D_1 and are diffracted less and reach the observer at O.

Rayleigh showed the degree of diffraction was inversely proportional to the fourth power of the wave-length of the light diffracted.

Since material particles consist of some sort of waves, may not similar phenomena occur when two particles meet? Niels Bohr suggested formulae analogous to Rayleigh's formulae describing the diffraction of light might be found to describe the interaction of particles consisting of wave-packets. He concluded that if the waves of a particle were very large they might not succeed in stopping, or interacting with, the waves of other particles which happened to be of smaller wave-length. By analogy with Rayleigh's wave-diffraction formula the probability of interaction between two sets of wave-packets would be inversely proportional to the square of the ratio of the lengths of the respective sets of waves. According to the modern wave-theory of matter the size of the waves in a particle's wave-packet is inversely proportional to the mass of the particle. Thus the probability of inter-action between two particles is proportional to the square of the ratio of their masses. The mass of the electron is nearly 2,000 times less than the mass of the neutron, hence the probability of interaction between an electron and a neutron is approximately $(1/2,000) \times (1/2,000)$, or 1 in 4,000,000 times less than for a neutron interacting with a nucleus. No wonder the experimenters could not find much evidence of inter-action between neutrons and electrons!

The existence of a neutron raises many possibilities. It has properties quite different from those of other particles of similar size, such as electrons and protons, because it has no electrical charge. The absence of an electric charge presumably should enable neutrons to approach each other to the point of contact, as neutral atoms are in contact in solids and liquids. The diameter of an atom is 2.10^{-11} cm. and the diameter of a neutron 2.10^{-16}. Hence, 100,000 neutrons would have to be laid in a row to equal the diameter of one atom. Hence $100,000 \times 100,000 \times 100,000$ could be packed into the space of one atom. This neutron package would constitute matter 10^{15} or 1,000 billion times denser than ordinary matter, say water. This enormously dense matter might stream through the material of any container. A quart pot of it would weigh 1,000 billion kilograms or approximately one billion tons, and yet it might stream through the spaces of the atoms in the material of the pot as water runs through a sieve, and leave the pot empty without crushing it. There is no evidence that the earth contains any super-dense neutron material. The stars may contain some of it. The very dense stars such as the white-dwarfs have densities sometimes a hundred thousand times that of water. According to E. A. Milne, all stars have an extremely dense, hot core. Perhaps neutron material may be shown to be there.

The neutron may prove to be a discovery of immense value to cosmology. The matter in the universe is supposed to have evolved through a series of states. All known matter consists of atoms made of bundles of protons and electrons; atoms of iron, oxygen, sulphur, etc. How were these bundles originally made? Were atoms made in the beginning of the universe, supposing there was a beginning, or have they been built up during the evolution of the universe? This seems probable, because atoms are known to be disintegrable; if they may be broken into bits presumably they may be built up again. The conception of the building of atoms of matter directly out of protons and electrons is very difficult, as these elementary particles are all electrically charged. Like electrical charges repel each other, so two electrons, or two protons, would not meet unless they approached each other with enormous velocities. As the nuclei of all atoms are positively charged, how can they be built out of an increasing number of protons? For instance, the nucleus of the helium atom, which has a charge of two positive units, contains two electrons and four protons. How could such a particle as the nucleus of a helium atom, which contains an excess of positive charges, accumulate more protons with positive charges to form the nucleus of the next simplest atom, that of lithium, which contains seven protons and four electrons? The construction of positively charged atomic nuclei such as those of iron, with a positive charge of twenty-six units, by the conglomeration of twenty-six and more positive units with some negative units, seems inconceivable. The building-together of an excess of twenty-six mutually repelling positively charged particles within a very small region cannot be imagined without extremely speculative reasoning. If they were built by the accidental simultaneous collision of fifty or more particles all moving at enormous speeds the laws of probability indicate that the construction of one of the larger atomic nuclei out of the primordial protons and electrons would be rare, and the existence of the countless millions of complicated atoms in the universe would be inexplicable without allowing the universe a gigantic age. The existence of neutrons eases this difficulty. A neutron would represent the first stage in the construction of compound matter. The combination of a proton and an electron is not difficult to conceive because their electric charges are opposite and opposite charges might conceivably attract each other and relatively easily come to stick together. The mechanism of the construction of a neutron out of a proton and an electron is not yet known. Some interesting speculations concerning its possible nature have been made by Langer and Rosen. They conceived the neutron as an abnormal form of the hydrogen atom. The normal hydrogen atom contains a proton as nucleus, with an electron revolving round it at a comparatively great distance. The exact distance depends on the state of energy of the atom. If it is highly charged with energy the revolving electron is in an orbit of greater diameter than if it is lowly charged. The neutron is conceived as a hydrogen atom in an abnormally low state of energy, with the proton and electron more or less in contact, or held together as the ends of a dumb-bell, or as a spherical object consisting of an electron with a proton at its centre. When an atom changes from a state of high to low energy it must emit energy during the process. It does so by emitting a quantum of radiation which may be in the form of a ray of visible light. The change of a hydrogen atom into a neutron would be accompanied, according to Langer and Rosen's speculation, by the emission of a large quantum of radiative energy, by a wave-radiation of great penetrative power. The still more recent discovery of the positive electron or positron has suggested that the proton may not be an elementary particle, but a compound of a neutron and a positron. If this is so, the neutron must be more elementary and primal than the proton.

Given the neutron, the origin of the complicated elements may be imagined as due to condensations of conglomerations of neutrons. As the neutrons have no electric field and do not repel each other they may easily come together in clumps. Suppose one of these clumps were struck by a fast particle from outside. Some of the neutrons might be disintegrated, and the now odd assortment of protons and electrons could rearrange itself into a collection of nuclei of atoms of any degree of complexity. Nuclei of atoms of iron, oxygen, magnesium, gold, and other elements would, as it were, crystallize out of the clump of neutrons. These are the atoms of more direct interest to humanity, as the materials of civilization are made out of them. The remarkable radioactive substances, though of immense scientific importance because of the knowledge they have provided concerning the

structure of matter, are not of much practical importance. They are extremely rare on the earth and have
a practical value limited by their expense and rarity.
They are useful in medicine; for producing luminous
watch dials visible in the dark; in certain delicate
methods of chemical analysis, and in other applications
which require small quantities of material only.
Many of the great achievements of recent atomic
physics have been made with the assistance of the
radioactive substances. The methods of research
have been dependent on them. One of the immediate
tasks of research is the liberation from the limitations
of radioactive techniques of investigation. The discovery of the neutron, and more particularly the
achievements of Cockcroft and Walton to be described presently, are the first attempts to solve the
chief problem confronting the second generation of
experimental physicists in the twentieth century.

Langer and Rosen have suggested that the cosmic
rays may be produced when clumps of neutrons are
condensed into the nuclei of helium atoms, by the
emission of some electrons.

Whether neutrons exist in the earth's atmosphere
is not yet decided. Their mass is equal to that of a
hydrogen atom. Hydrogen atoms are not found in
the atmosphere because they are very light and ascend
to the top, whence they escape. Their speed of
motion under atmospheric conditions is sufficient to
enable them to overcome the earth's gravitational
attraction. Hydrogen is always being lost from the
earth into outer space. At the first consideration the
same causes might effect the escape of any neutrons
in the earth's atmosphere into outer space. At the
second consideration this is not certain. Other
factors besides mass affect the escape of atoms from
the earth. The mean free path, that is, the average

distance a particle travels without collision, is also a factor. As neutrons have immensely long free paths, owing to their great penetrating power, they may behave differently in this circumstance from hydrogen atoms. The escape of neutrons from the top of the earth's atmosphere is not certain and will have to receive further scientific study before it can be decided.

The extreme interest of the neutron for the experimental physicist is due to its provision of a new technical agent in research. The nucleus of the atom of helium has hitherto been the most important particulate agent. The relatively high mass and speed of these particles ejected from radioactive substances has endowed them with unequalled disintegrative power. The differences between the helium nuclei and electrons enabled many fundamental experiments to be made which could not be done with electrons because of their much smaller mass. The differences between neutrons and helium nuclei may similarly bring a series of new phenomena within the knowledge of physicists. The peculiar properties of the neutron are as yet little known, and provide a strong stimulation for study. Their unknown properties may when discovered reveal unsuspected aspects of nature. This is the explanation of the immediacy of interest in the neutron shown by such a master of the scientific imagination as Niels Bohr. Rutherford has said that the discovery of X-rays in 1895 gave an unequalled stimulus to physical research. They were a new and unsuspected natural phenomenon. Their appearance broke the hardening circle of the physical ideas of the nineteenth century, and shocked the experimentalist into the physical ideas of the twentieth century. The discovery of the neutron may provide a similar stimulus during the coming decades.

II

The first artificial disintegration of atoms without the assistance of radioactive substances is another recent achievement of workers in the Cavendish Laboratory. It marks the emancipation of experimental atomic research from the limitations of rarity and lack of variety imposed by the use of radioactive substances. The available quantities of radium and its fellows are small, so the experimenters cannot obtain intense beams of rays from them. The number of particles emitted by a few thousandths of a grain of radioactive powder is comparatively small, and cannot be increased because larger quantities of the powder are not available. The number of known naturally radioactive substances is small and their properties are definite, so the experimenter has a limited choice of rays from them. Some emit one sort of ray only, which in the case of polonium is helium nuclei. Others emit electrons only, and others a mixture of two or three usual sorts of radiation, i.e., of electrons, helium nuclei and wave-radiation. The emitted rays have a narrow range of properties, as the particles are restricted to definite speeds, and the wave-radiations usually to waves of a certain length. The experimenter must work with the limited range of speeds and sizes provided by the naturally-occurring radioactive substances. He is under handicaps similar to those of a motor-driver whose gear-box is restricted to a few gears, and of a general whose artillery is restricted in numbers and calibre. The atomic investigator requires an infinitely variable number of calibres of atomic batteries. Some of the methods of making such an artillery are simple in principle. Atomic particles may be caused to move by exposing them to an electric force. Their speed will depend on the

PLATE II

Dr. J. D. Cockcroft and the apparatus with which he accomplished, in collaboration with Dr. E. T. S. Walton, the first disintegration of an atom by a machine. Previously, the artificial disintegration of atoms had been accomplished only with the assistance of radioactive substances. (L.E.A.)

magnitude of the electric force applied to them, so particles of almost any speed may be produced by varying the strength of the applied electric force. Just build a big machine giving a comparatively strong electric current at high voltage and submit a stream of particles to it, and you will have the atomic particles in any number and of any speed you wish. This straight-forward method is beset with technical engineering difficulties. The nuclei of atoms are a billionth of an inch in diameter, and a powerful electrical machine may be as big as a house. The big machine may easily prove too clumsy to be effective. It may produce the streams of high-speed particles but not manageably. If the machine will not respond perfectly to control and will not act with precision it will not give precise effects the observer can measure. The production of brilliant, but ill-defined efflorescences or photographs yields little information. The speed of particles may be increased by impulse : short and sharp acceleration, or by cumulative methods.

The American physicist, E. O. Lawrence, of Berkeley University, has succeeded in constructing a cumulative impulse machine. Cockcroft and Walton obtained swift particles by submitting particles to the short and sharp pressure of one powerful electric field. Lawrence has obtained swift particles by submitting particles moving in a circular or spiral track to repeated impulses. His apparatus is in principle an atomic electric motor, the armature of which consists of particles restrained to a circular track by a magnetic field, and impelled round this track by repeated electric pushes. With quite small powers he can gradually accelerate protons and other particles up to an energy of several million volts. He can achieve this result with small power because his method is cumulative or resonant, and, as in most motor engines, produces a powerful result by the repeated application of comparatively small forces. His apparatus consists essentially of two plates A and B, as in Fig. 6.

A high potential difference is produced between the plates. As these are of metal and conductive, the electric field will not penetrate into the regions D, E. But these regions may be subjected to a circular magnetic field. Suppose a particle is at P, and is impelled by the electric field between A and B towards B. If

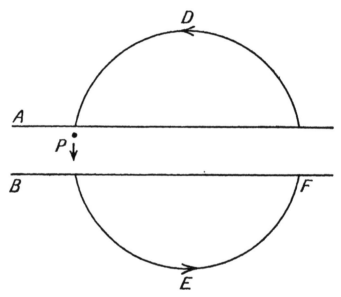

FIG. 6.   The principle of Lawrence's apparatus.

there is a hole in B it will rush through and be whirled round the circular magnetic field until it hits the plate at F. If there is a hole at F it will enter the field between B and A. Suppose the direction of the field has been changed during the time the particle travelled from P to F. Then the particle would be accelerated again as it passed from F through B towards A. Hence a suitable system of holes in plates A and B, and an alternating electric field between the plates should theoretically allow the production of repeatedly accelerated particles. The practical difficulties in

Professor E. O. Lawrence and his atomic motor for producing swift particles by cumulative impulses. He applies a pressure of a few thousand volts repeatedly until particles of millions of electron-volts energy are obtained. He made the first atomic disintegrations with diplogen. (E. L. Lawrence.)

[face page 30

constructing such an apparatus appear in the first consideration to be insurmountable, but Lawrence has overcome them and obtained supplies of swift particles of an energy of several million volts. His apparatus promises to provide streams of particles of 15,000,000 volts energy. His successful application of the resonance principle of all rotating motors is a great experimental achievement. He has already made remarkable discoveries by bombarding various atoms with heavy hydrogen nuclei of mass 2. The introduction of rotatory movement into machinery usually marks a fundamental improvement in design, and Lawrence has accomplished this. He has constructed the first atomic motor.

In his first attempt to produce swift particles by extra accelerating forces Cockcroft tried to use a method similar to Lawrence's, but he abandoned it owing to the practical difficulties in making it work. He decided to attempt to develop the straightforward impulse method. When a strong electric force is applied to the terminals of a tube containing gas at low pressure a discharge occurs and a current passes through the tube. It passes because the electrical pressure has caused many of the atoms of gas to become ionized and lose electrons, so that many positively and negatively charged particles are liberated. Under the impulsion of the electric field the positively charged particles, i.e., the normal atoms minus an electron, move towards one terminal and the detached electrons to the other terminal. The movement of these charged particles constitutes the electric current. If the gas in the tube is hydrogen, the ionized hydrogen atoms are split into their two components, each liberating one proton and one electron. The speed of the opposing streams of protons and electrons depends on the electrical pressure, the voltage, applied to the tube.   By applying hundreds of thousands or millions of volts very swift streams of particles may be produced inside the tube.

In Fig. 7 the concentric rings represent normal hydrogen atoms of one proton (small circle) with one revolving electron (small dot).   The small circles or protons are streaming to the negative terminal, the small dots or electrons to the positive terminal.

A tube in which experiments may be conveniently made is not easily designed and worked.   The designer must also produce powerful and constant electrical fields.   These technical difficulties cannot be solved without the highest engineering skill.   This point raises the question of the dependence of pro-

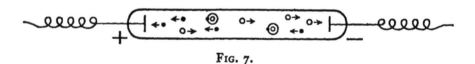

FIG. 7.

gress in physics on the utilization of developments in engineering.   As engineering is stimulated by industrial requirements, and industry is stimulated by the struggles of industrialists for commercial gain, there is a connection between the progress of physics and the forces of social competition.

The influence of social forces on the design of Cockcroft and Walton's apparatus may clearly be seen. The experimenters obtain the swift protons from a tube containing hydrogen, on the principle shown in Fig. 7.   The protons are submitted to a powerful acceleration as they pass down the tall six-foot vertical glass tube G shown in Fig. 8.   The protons are released in the small discharge tube at the top and pass through the water-cooled canal into the big accelerating tube.   The big and little tubes contain many joints, they would be extremely difficult to

manipulate without joints because the best method
of arranging the apparatus could not be discovered

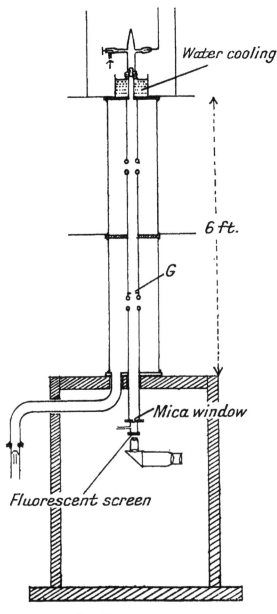

FIG. 8. (Cockcroft and Walton.)

without trying the parts in various positions. These
large and small tubes must nevertheless be capable
of keeping a good vacuum, if protons are to be

produced and accelerated in them.   How could this necessary good vacuum be kept in an apparatus containing so many joints ?   It could be kept by making most of the joints out of fused glass, the whole six-foot erection might be fused into one continuous glass structure.   If that were done the apparatus would become experimentally unmanageable, because glass joints cannot easily be broken and mended.   Once the apparatus had been arranged it would have had to remain so.   The experimenter would have been unable to arrange the glass parts of his apparatus, his hands would have been tied.   A flexible method of making air-tight joints is of first importance in the design of an apparatus such as Cockcroft and Walton's. The substance used for making the joints is plasticine, the well-known children's modelling material.   The discovery that plasticine will make good vacuum joints has had an important influence on the progress of physics.   Besides good and easily made joints the experimenter requires an easily kept vacuum, he must have powerful vacuum pumps which will rapidly reduce and keep the pressure inside the apparatus suitably low.   If the vacuum pumps take a week to exhaust the apparatus the experimenter's initiative is much restricted.   He cannot afford to alter the apparatus and break the air-tight joints frequently because he would have to wait a week before he could obtain any more fast protons for experiments.   If the vacuum pumps are powerful and efficient enough to reduce the pressure in the apparatus from atmospheric to the working low pressure in a few minutes, the experimenter has enormously more control over his apparatus.   He may make dozens of experiments when otherwise he could make only one.   The experimenter requires good joint-making substances and efficient pumps.   Some of these technical necessities have been provided through the assistance of the remarkable plastics and oils prepared by Dr. C. R. Burch. Dr. Burch was on the research staff of a large electrical manufacturing company. The large electrical firms of the world are connected in an international financial trust. The directors of this trust seek the usual social satisfactions of power and wealth by causing the communities to become more and more dependent on the products of electrical engineering. As the world consumption of electrical goods increases the power and wealth of the electrical companies increases. The world trustification is of a loose structure, allowing competition and emulation between the American, German and British and other sections of the trust. In the course of emulation the various companies and their research staffs make improvements which will enable some to lead for a time, and then others, as in a league of sporting clubs.

Improvements in fundamental manufacturing processes have much economic importance. For instance, methods of increasing the efficiency of the processes for manufacturing gas-filled electric lamps, and radio valves. The world consumption of gas-filled lamps is of the order 1,000,000,000 per annum. Improvements in pumps which will enable lamp and radio valve glass bulbs to be exhausted with greater efficiency, in materials which make good metal and glass joints such as are necessary in lamps and valves, have immense economic significance. The struggles of the financiers for control of the electrical companies and their struggles with the rest of the people for the maximum power and profit from their companies create the active atmosphere in which talented scientific technologists are stimulated to discover useful improvements. Dr. Burch discovered some remarkable oils and greases which evaporate with extreme slowness.   These are suitable for lubricating the vacuum pumps, as, unlike other oils, they do not emit appreciable quantities of gas into the apparatus. Cockcroft and Walton's apparatus could not have been worked without using Burch's oils.   Thus the most important recent experiment in atomic research has direct relations with technological development, and through that with competitions between financiers, and between financiers as a class and the rest of the people.   The relations are exemplified in other parts of the apparatus.   ·When the protons enter the tube G they come under the influence of an electrical field which may be increased up to 800,000 volts.   The electrical potential is applied to the tube G through a system of transformers, rectifiers and condensers (Fig. 9).   All of these machines are products of modern engineering practice.   Electrical energy is most conveniently distributed as high-voltage alternating current, but users require low-voltage and often direct current, so transformers to transform the voltage, and rectifiers to convert alternating into direct current, are important industrial products.   As the size of electrical power stations, and the range of distribution of electricity, increases, the size and efficiency of transformers and rectifiers must be increased. High voltages require good insulation.   Improvements in insulation enable more compact and therefore more manageable apparatus to be constructed.   Cockcroft used transformers of modern design in his apparatus.   He drew deeply on the resources of the most modern electrical engineering.   His arrangement of transformers, condensers and rectifiers allowed a steady potential difference of a value up to 800,000 volts to be applied to the ends of the tube G.   The protons entering this field from the small discharge tube at the top were accelerated

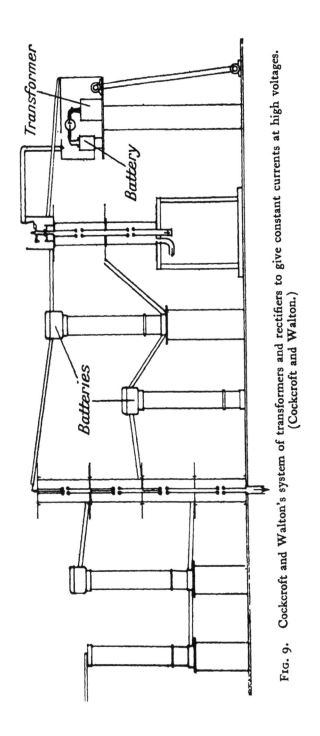

FIG. 9. Cockcroft and Walton's system of transformers and rectifiers to give constant currents at high voltages. (Cockcroft and Walton.)

down the tube. At the bottom of the tube various objects could be exposed to the descending stream of swift protons. The observer seated in the chamber could watch through the microscope the effect of the impact on specimens of various substances exposed to the proton stream. The arrangement of the target bearing the exposed specimens is shown in Fig. 10.

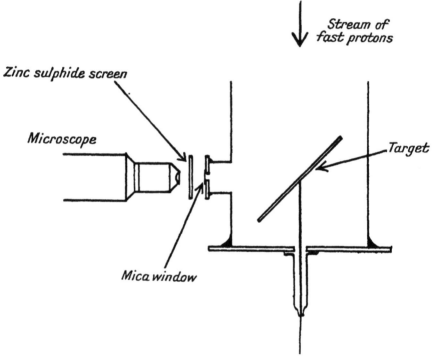

FIG. 10. (Cockcroft and Walton.)

By placing the mica observation window at right angles to the direction of the stream, and at a sufficient distance, protons are prevented from straying sideways and passing through the mica window. The possibility of protons from the stream straying through the mica may be prevented also by choosing a suitable thickness of mica. The speed of the protons down the tube is known and their penetrating power through mica may be deduced from their speed. If the mica is of a thickness just too great to allow the penetration of protons of the maximum speed producible by the apparatus, no protons could escape through the window. Any particles escaping through the window must be of different origin.

Cockcroft and Walton discovered that if the target was made of the light metal lithium particles did escape through the mica window when the lithium was submitted to the impact of the stream of swift protons. As the particles could not have come from the proton stream they must have come from the lithium. This was very sure because particles were detected outside the mica window when the protons were moving under the surprisingly low electrical field of 125,000 volts. Apparently the lithium atoms had been disintegrated by relatively slow protons. The disintegration of atoms by particles from radioactive substances had been accomplished only with extremely energetic helium nuclei whose energy was of the order several million electron volts. Cockcroft and Walton expected a field much stronger than 125,000 volts would be necessary. If they had had no guidance except from the disintegrations with radioactive substances they would have expected to have had to use fields of several million volts to give the protons sufficient energy to be able to disintegrate lithium atoms, but recent theoretical researches had shown that disintegrations might occur at voltages much lower than these. Without the support of the theoretical researches of Gamow, and Gurney and Condon, concerning the explanation of the mechanism of the disintegration of radioactive atomic nuclei, Cockcroft and Walton could not have had sufficient confidence in the effectiveness of their apparatus at low voltages. A sketch of the ideas of Gamow, and Gurney and Condon, will be given presently.

The passage of particles through the mica window

was demonstrated by several methods. If a screen covered with zinc sulphide is placed opposite the window, swift atomic particles impinging on it cause little green scintillations. An experienced observer can deduce the sort of particle from the sort of scintillation produced. Electrons, protons and helium atomic nuclei make rather characteristic scintillations of different sizes, according to their differences in mass. The massive helium nuclei make scintillations larger than those produced by the less massive protons.

When the lithium was bombarded with 125,000 volt protons scintillations were observed on the zinc sulphide screen. They appeared to be due to helium nuclei. Helium nuclei had been emitted by lithium atoms disintegrated by the swift protons. Lord Rutherford has given an already famous description of the occasion of these great experiments, the first disintegration of atoms by humanly constructed machinery. When he inspected the scintillations he recognized that they were due to helium nuclei. The existence of swift rays of helium nuclei or particles had been, thirty years before, one of the earliest of his great discoveries. When he examined the scintillations caused by the particles from the lithium he at once recognized, he said, the nature of the particles, for he had been present at the birth of the α-particle and he instantly recognized that these were legitimate children.

The range and penetrating power of these particles from lithium indicated that they were nuclei of helium atoms travelling with an energy of about 8,000,000 electron-volts. They were very much more energetic than the swift protons which had released them from the lithium atomic nuclei. Evidently the protons were acting as atomic triggers or detonators, they were causing atomic explosions much more violent than themselves.

Were these results to be expected? The lithium nucleus contains seven protons and four electrons. If it captured a proton the nuclei group would contain eight protons and four electrons. These most reasonably might split into a pair of helium nuclei or α-particles each of which contains four protons and two electrons.

The production of a pair of helium nuclei from the combination of a lithium nucleus and a proton was reasonable. But if the helium nuclei were produced in pairs they should be detectable in pairs. As the

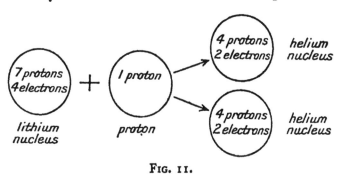

FIG. 11.

energy of motion of the helium nuclei is enormously greater than that of the incident proton which helped to produce them, it must come from within the lithium nucleus. Hence the total energy of the pair of helium nuclei must be constant in any direction. As they are formed by an internal atomic explosion they must fly off with equal speed in exactly opposite directions. If that were not so, energy and momentum must have entered from outside the system, and the energy of the incident proton was known to be relatively negligible. Cockcroft and Walton immediately devised an experiment to discover whether the helium nuclei were emitted in pairs whose members were travelling with equal speed in opposite directions.

Screens were placed at equal distances on opposite sides of a specimen of lithium (Fig. 12). The high percentage of simultaneous impacts on the two screens indicated that the helium nuclei were probably being emitted in pairs. Subsequently, P. I. Dee and E. T. S. Walton obtained beautiful Wilson chamber photographs of the twin disintegrations, examples of which are reproduced. As each helium nucleus had an energy of approximately 8,000,000 electron volts, the total energy released in the lithium nucleus explosion was 16,000,000 electron-volts. This suggested several interesting considerations. Was this output of energy in a lithium atomic explosion of a probable order?

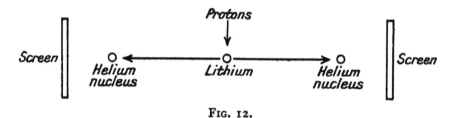

FIG. 12.

The masses of lithium and helium nuclei and of protons are not exactly whole numbers in terms of each other. The mass of helium nuclei is 4·0011, and that of a lithium nucleus is 7·0104, and of a proton 1·0072 units. When the lithium nucleus and the proton react together the mass changes may be represented:

Lithium   Proton = Helium + Helium + free mass
7·0104 + 1·0072   4·0011   4·0011   0·0154

The sum of the masses of the pair of helium nuclei is slightly less than that of the sum of the masses of the lithium nucleus and the proton. What happens to the 0·0154 units of mass that have been left over? They are converted into energy, and this energy gives the immense speeds to each of the helium nuclei.

PLATE IV

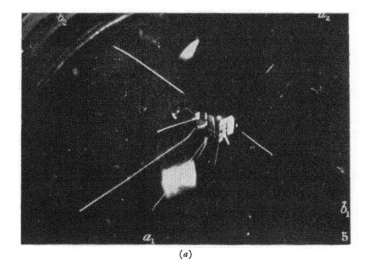

(a)

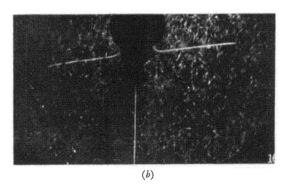

(b)

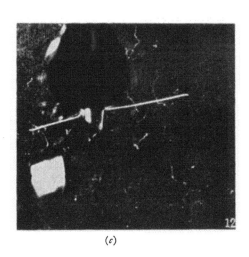

(c)

(d)

(a) The disintegration of atoms of lithium by swift protons. The pairs of tracks a_1 a_2, b_1 b_2, are due to pairs of helium atoms. These have been formed by the combination of a helium atom with an impinging proton, followed by the splitting of the combination into a pair of helium atoms.

(b) The disintegration of an atom of boron by a swift proton. Boron is believed to absorb the proton and split into three helium atoms. In this case, two of the helium atoms may have got most of the energy and the third has not enough to appear. The third track is probably not connected with the pair of opposite tracks.

(c) Disintegration of an atom of lithium by an impinging atom of diplogen or heavy hydrogen.

(d) Disintegration of an atom of lithium by an impinging atom of diplogen with the ejection of a pair of helium atoms. A third fine track also appears, probably due to an atom of ordinary hydrogen.

(P. I. Dee and E. T. S. Walton, and Royal Society.)

Einstein has shown that mass is a form of energy, a congealment of it. All the matter in the universe is congealed or tied energy. Untie it, and it may be passed to the particles around so that they are flung forth. Cockcroft and Walton's experiments show how atomic mass may be transformed into energy through the agency of a humanly constructed machine. The amount of energy so released is absolutely very small. At 500,000 volts only one out of every thousand million protons accelerated succeeds in causing a disintegration. Besides, only a small part of the electrical energy needed to operate the machine is communicated to the protons. There is an immense loss of energy in the accelerating of the protons, and again in the number of disintegrations per proton, so though the energy released in one atomic disintegration is much greater than that of the impinging proton, it is nevertheless immensely less than the energy needed to operate the whole apparatus. In the present form the apparatus does not offer a solution of the problem of the production of energy from mass for industrial consumption. Mass is an exceedingly concentrated form of energy. The material of a chair or a brick contains enough energy to drive a liner across the Atlantic Ocean. If the mass of a brick could be annihilated and transformed into energy there would be sufficient to supply the electrical demands of London for one week. The atoms of all the matter in the universe are packets of energy. We do not know why the universe is stocked with this energetic stuff. It is a fortunate circumstance because the atomic energy is much more accessible than might have been expected. If atomic energy, mass, were bound in more stable packets it might for ever have been beyond human research. The matter of the universe is not infinitely stable.   If it were, the evolution of the universe would be more than ever unintelligible.   The existence of matter, the earth, sun and stars, which have the marks of an evolutionary history, suggests that matter may not be so difficultly formed out of energy.   If the packets of energy, in the form of atoms, exist, they must presumably have been made up some time, and therefore presumably be untieable again.

Humanity has yet to discover how the explosive unstable atoms of matter may be touched off in any more efficient way.   The quantity of energy required to detonate matter may be much less than the quantity of energy so released, as the quantity of energy needed to detonate a charge of dynamite is much less than the energy liberated in the explosion.

While the apparatus and experiments of Cockcroft and Walton direct attention to the accessibility of atomic energy, they do not as yet suggest any method of obtaining atomic energy efficiently.   A practical atomic energy generator must release more energy than is needed to operate it.   Cockcroft and Walton's apparatus works with an efficiency of perhaps one hundredth of one per cent.   Theoretically it probably could never work with an efficiency of more than four per cent; i.e. it could not theoretically produce a quantity of atomic energy equal to more than one twenty-fifth of the electrical energy needed to operate it.   If the incident protons were given an energy of 500,000,000 electron-volts, the percentage that accomplished an atomic disintegration in a foot-cube of lithium might be large, and the amount of atomic energy released would be large, but the energy needed to produce the 500,000,000 volt protons would be much larger.   These considerations make the physicists engaged in disintegratory researches deplore the raising of premature hopes of providing new sources of energy for industrial and social use. But it is difficult to believe that the recent success in releasing some atomic energy by humanly constructed machinery will not ultimately, if only by entirely different methods and centuries hence, lead to the discovery of efficient methods of releasing atomic energy, in which the amount of energy released will be many times greater than the amount of energy needed to operate the machinery. The neutron is the present slender hope for the production of atomic energy for practical use. It interacts only with atomic nuclei which contain the energy to be released. It exerts its energy on breaking into the atomic power-house, and is not exhausted in friction with the clouds of electronic sentinels surrounding the nuclei.

Cockcroft and Walton have found that the swift protons disintegrate other atoms besides those of lithium. Beryllium gives a few helium nuclei, not observable outside the tube under voltages of 500,000. Boron breaks up at the low voltage of 115,000. It may split into three helium nuclei according to the formula:

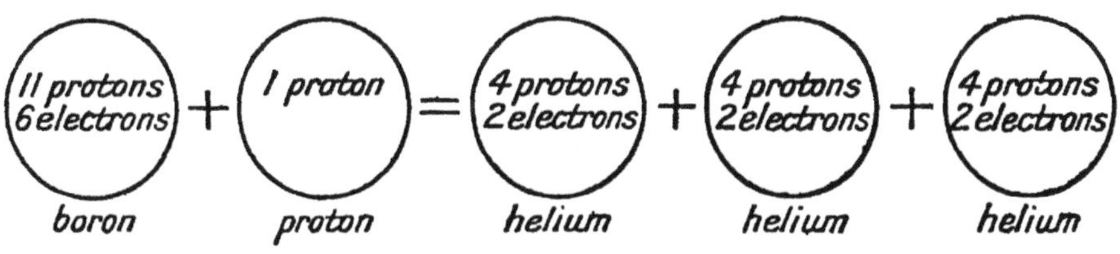

$$\left(\begin{array}{c}11\,protons\\6\,electrons\end{array}\right) + \left(1\,proton\right) = \left(\begin{array}{c}4\,protons\\2\,electrons\end{array}\right) + \left(\begin{array}{c}4\,protons\\2\,electrons\end{array}\right) + \left(\begin{array}{c}4\,protons\\2\,electrons\end{array}\right)$$

boron proton helium helium helium

FIG. 13.

Cockcroft was educated and trained as an electrical engineer. After working in industry for two years as a fully qualified electrical engineer he was awarded a scholarship which enabled him to study applied mathematics at Cambridge. When he started research in physics he was comparatively old, as he was then nearly thirty. The sort of training becoming necessary in some branches in experimental physics is excellently illustrated by his history. First, a thorough school and university course ; second, some years as a practical engineer ; third, further study in mathematics. This is the sort of equipment now necessary to the physicist, who must understand modern mathematical physics and possess the engineering skill and experience necessary in the design of powerful machinery. Similar qualifications are noticed in Professor Kapitza, who has devised the famous machinery for producing super-magnetic fields. The physicist must understand the mathematics which discovers to him the significant points in the new theories. If he does not have a general understanding of these theories he cannot be inspired by them to make illuminating experiments, and if he has not the engineering skill and experience he cannot devise the powerful machines necessary to test many points in these theories.

The contributions of electrical engineers to recent physics are remarkable. Einstein was trained as an electrical engineer, and the leading British theoretical physicist of the day, Dirac, was also trained as an electrical engineer. Does this indicate that the engineering imagination is of special value to the contemporary theoretical physicist? That would be a deduction disagreeable to certain fashionable opinions, as the mechanistic mode of theorizing, and describing nature, is supposed to have become of little use since the introduction of quantum mechanics. May it not be contended that quantum mechanics is just as mechanical as any other sort of mechanics, and in so far as it describes phenomena, the world is susceptible of mechanistic description? In fact, quantum mechanics has described the world more successfully than any other theory ; may not quantum mechanics therefore be held as proof of the success of the mechanistic description of nature? Dirac writes that recent advances in theoretical physics imply "a lack of arbitrariness in the ways of nature". Is not lack of arbitrariness the essence of the mechanistic philosophy of science?

The theoretical researches of Gurney and Condon, and Gamow, published in 1928, gave evidence for the possibility of disintegration of atoms by protons travelling under comparatively low voltages. The helium nuclei emitted by radioactive atoms have an energy of several million electron-volts. This had suggested that an atomic projectile of equal energy might disintegrate an atom, and Rutherford had shown that that can be arranged experimentally. The argument is of

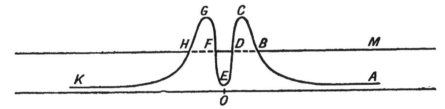

FIG. 14. (After Gurney and Condon.)

common sense and not based on an understanding of the nature of the mechanism of radioactive disintegration. The first insight into the mechanism of natural disintegration was published by Gurney and Condon. They applied the new conceptions of the wave-theory of matter to the constituents of the nucleus.

Suppose Fig. 14 represents a cross-section through the nucleus of a radioactive atom. The centre of the nucleus is at O. The wavy line represents the strength of the forces at different distances from the centre of the nucleus. The forces are low at the centre O, as the curve falls down to E. At short distances from the centre the forces have risen to high values G and C. At further distances the forces diminish rapidly.

Suppose that the line HFDBM represents a force which cannot be overcome by a particle of energy less than 4,000,000 electron-volts. Then another horizontal line through GC would represent a force which would control a particle of 8,000,000 electron-volts, as it is twice as far as HFDBM from O.

Suppose a particle within the nucleus is moving with an energy of 4,000,000 volts. Then it will never, as it were, rise above FD. It will be shaking within the area FED. Its behaviour will resemble that of a ball-bearing within a channel or valley. If the ball is held stationary at B, as in Fig. 15, and then dropped, it will slide down to the bottom and run up the other side to an equal height, and back again. It

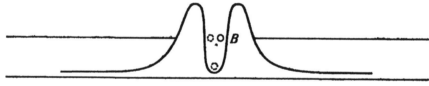

FIG. 15.

will never get out of the valley if it was started at a level B inside. If atomic particles were as steel balls radioactivity would be incomprehensible. According to the wave-theory of matter, particles are systems of waves or behave as if controlled by waves and not strictly defined objects with sharp surfaces. The ball-bearing ought to be replaced by a wavy structure whose core is a concentrated pocket of waves, and whose outer fringe fades away rapidly. The more correct representation should be therefore as in Fig. 16. The circles represent wave-fronts of probability, the probability that the energy of the particle will appear as if concentrated at that place at any moment. The distance between the successive circles is a measure of the improbability of the presence of the concentration. According to the wave-theory of particles, therefore, the energy of the particle will nearly always be inside the valley, but sometimes it may appear outside if probability indicates that it should be on one of the outer circles. The energy of the particle and the size of the valley of forces inside the nucleus may be such that the waves of probability guiding the apparition of the particle may exist in considerable strength outside the valley. Then the particle will sometimes appear outside the valley and escape altogether. It will escape with the energy it possessed while inside the valley. The probability of disintegration depends solely on the size of the valley and the energy of the particle inhabitants. In the

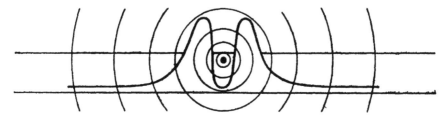

FIG. 16.

majority of atoms the wave-front of probability outside the nuclear valley will be extremely faint, and the chance of the particle appearing outside will be extremely small. In an atom of uranium the outer waves of the particle which stretch beyond the valley are occupied on an average only once in five billion years, whereas in Radium C^1, a large section of the particle's probability waves stretches outside the valley, so that the energy of the particle may reside outside the valley extremely frequently, in fact, once in a millionth of a second. In stable elements such as iron the valley of forces in the nucleus and the energy of the contained particles is such that the energy of the particles virtually never appears in the faint distant waves beyond the sides of the valley. According to the wave-theory of matter

E

all atoms containing an α-particle with energy enough to escape must be more or less radioactive because the energy of a particle may appear at any moment anywhere in the universe. 'But the probability of its energy appearing more than a millionth of an inch from where it was at the previous moment is in general small beyond imagination. Atoms are radioactive because the position of a particle in the universe is determined by chance. The particle may be here at one moment and in the moon at the next. The improbability that it will appear at such a great distance from its previous place is immense, but the improbability that it will appear one billionth of an inch outside an atomic nucleus after being inside it at the previous moment may for atoms of suitable structure not be great. Radioactivity is a consequence of the statistical nature of the universe; it follows from Heisenberg's Principle of Uncertainty. No particle occupies an absolutely precise place in the universe. The position of a particle cannot conceivably be measured beyond a known limit of accuracy, and this limit proves to be of the same order as the sizes of atomic nuclei. The relative sizes and states of the nuclei and their constituents in some atoms are such that the perceptible vagueness of the position of some of the constituents extends to the region outside the nucleus. Radioactive disintegration is not explosive in mechanism. As Gurney and Condon wrote, the α-particle slips away from the radioactive nucleus almost unnoticed. The waves of probability of position of particles obey the same mathematical laws as other waves. The aspects of particle-waves which account for radioactive disintegration may be paralleled by aspects of other wave-phenomena. For instance, the phenomenon of the total reflection of light at the surface between two transparent substances is well known. Many must have seen a ray of light directed through a trough of water to be reflected on the inside of the water surface.

If the incidence of the ray on the water surface is acute the ray does not pass into the air, though both water and air are transparent, but is totally reflected at the surface, as shown in Fig. 17. If light consisted of small particles they would be redirected at A without penetrating the air, as if they were billiard balls and the water-air surface were a cushion. As the ray consists of waves the phenomenon is much more complicated. The disturbance at A does penetrate

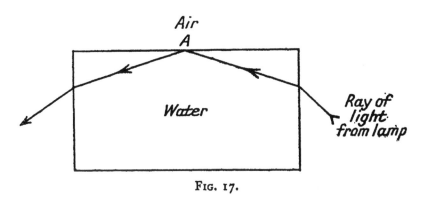

FIG. 17.

a short distance into the air, and the law of the decay of its intensity is the same as the law of the decay of the intensity of the probability of a nuclear particle appearing outside its nucleus. Another example of the analogies between the wave-theory of light and of particles was given in the section describing Niels Bohr's explanation of the rarity of interaction between electrons and neutrons when they collide. In that case, the phenomenon of the diffraction of light was remembered; in the case of radioactive disintegration, the phenomenon of the total reflection of light at the surface between two transparent substances.

Gamow made approximate calculations of the strength of waves of various nuclear particles outside their nuclei. He discovered that it was in some cases unexpectedly large. This implied that a particle of comparatively low energy approaching the nucleus from outside might get inside; as the approaching particle's waves would have the same power of penetrating the energy walls of the nucleus as particles of equal energy inside could penetrate outside. This discovery gave Cockcroft reason to believe that particles of moderate energy might penetrate the nuclei of some sorts of atoms.

In many of the disintegrations of atomic nuclei by bombardment with α-particles from radioactive substances a proton is ejected. For instance, a nitrogen nucleus emits a proton when struck by a fast α-particle. Blackett showed that the nitrogen nucleus captured the on-coming α-particle and did not break up into other bits. Cockcroft finds bombardment with protons produces in a number of cases α-particles. So the phenomena he is investigating seem to some extent to be the reverse of the mechanism of artificial disintegration with radioactive projectiles. This relation between the first method of atomic disintegration involving the use of radioactive substances and the first method of disintegration by machinery is striking. In the first method the α-particle goes in and a proton comes out; in the second a proton goes in and an α-particle comes out. The relation between the two phenomena increases the comprehensibility of the atom, for if they were apparently unconnected they would suggest the possibility of a much larger variety of disintegratory mechanisms, and an excessive extension of the regions of human ignorance.

The new examples of atomic disintegration found by Cockcroft and Walton have a bearing on the constitution of the stars and the history of the universe. The relative quantities of elements in the earth and stars must be connected with the ease of disintegration. Elements easily disintegrated by protons should be rare, and so the rarity of lithium is not surprising. The cores of stars contain protons and electrons shooting about at high speed due to the high temperature. How could lithium atoms escape disintegration? Only if the temperature were sufficiently low. The existence and relative quantities of the proton-disintegrable elements in the atmosphere of stars, which may be determined spectroscopically, may provide interesting new knowledge concerning the age and nature of the stars. As Eddington has elegantly explained, the studies of astronomy and physics advance in unison, the structure of the atoms elucidating the constitution of stars, and the behaviour of stars the properties of atoms.

CHAPTER II

DER KOPENHAGENER GEIST DER QUANTENTHEORIE

I

FOR the last fifty years the Cavendish Laboratory has been the leading school in the study of experimental atomic physics. More important experimental investigations of atomic phenomena have been made by students of this school than by those of any other experimental school. The flair for experiment seems to be the most powerful aspect of the British scientific genius, and consists in a combination of manual dexterity with theoretical understanding. Successful experiments are not made by just trying things, they are achieved consistently only when the manual operations are severely disciplined by the theoretical understanding. Empiricism is not the soul of scientific experimentation, though an important element. In a sense, experimental science is concerned with material more intractable than that of theoretical science. An apparatus is not as fluid as a mathematical symbolism. It appeals, therefore, particularly to those with a strong individualistic tendency, because method in experimentation is much less complete though no less important than in theoretical researches assisted by the study of symbolisms written on paper. A mathematical physicist needs no apparatus other than pencil, paper and library, which are of the most tractable nature. As the material he works with is tractable, his work is susceptible of a completeness of method which appeals to those with well-disciplined intellectual habits. Continental students receive an intellectual discipline more thorough than is customary in Britain. This may help to explain the higher average level of Continental theoretical researches. The number of geniuses produced by Germany and by Britain is proportionately equal, but British theoretical workers of the third rank fall far below the German in quality and numbers. The excellence of the German theoretical training accounts for the large number of dull German scientists, often criticized by the British and supposed by them sometimes to indicate an essential dullness of the German intelligence. The explanation is to Germany's credit. In Britain this number of dull researchers is smaller as they drop research sooner, partly because teaching posts are more easily obtained here. The British tradition of considering nothing worth doing unless done well is aristocratic. The history of science in Britain shows the profound influence of wealthy leisured investigators determined in their choice of subject and method by social habits and outlook. A taste for experimental research may have an aristocratic factor in its constitution because experiments cannot be made without wealth, and the community does not offer wealth for experimental investigation until individuals have shown their work might be worth subsidizing. The distinction of British investigators in experimental physics may be partly due to a tradition born in the capitalistic form of society which developed in Britain earlier than in other countries. The early British experimentalists were usually rich. They aristocratically liked to do as they pleased, and their work had an originality which came partly from the intellectual independence

in which they could indulge because they were econo-
mically independent.   While economic independence
gave them freedom of imagination it also made them
impatient of intellectual discipline.   They could afford
to leave experiments unfinished because they were not
being paid to finish them.   The absence of necessity
for continuous work disinclined them for the drudgery
of thorough theoretical training.   The thoroughly-
trained theorists could not have been expected to have
evolved out of the rich English gentlemen-experi-
menters ;   they had to be evolved out of a different
social class.

As the British leadership in experimental atomic
physics has been partly determined by British social
history, the Continental leadership in theoretical
atomic physics has been partly determined by Continental social history.   The prestige of the professional
classes in relation to the capitalist classes in Continental
countries is greater than in Britain, so the cultural
tradition of the professional man is there more dominant.   This tradition is one of thorough theoretical
and book-learning, and causes the student to be disposed more to the study of theory than of experiment,
to the spirit of the lawyer rather than that of the entre-
preneur.   The Continental scientists have led, especially recently, in the attempts to give orderly accounts
of natural phenomena.   In the realm of the theory
of atomic physics the leadership has for the last decade
resided in Copenhagen.   Niels Bohr is the presiding
genius.   For a number of years now the chief young
theoretical atomic physicists have gone to Copenhagen
to join his school of theoretical research, as soon as
they have exhibited notable ability.   The intellectual
soil of the Institute for Theoretical Physics causes
growing ability to mature in splendid forms.   The
two most active theoretical researchers of the day

PLATE V

(*a*) The Institute for Theoretical Physics, Copenhagen. (N. Bohr.)

(*b*) The Physico-Technical Institute, Kharkov. (See page 80.) (I. Obreimov.)

are Heisenberg and Dirac. Their genius was not fully developed until they had experienced the sympathy and understanding of the "*Kopenhagener Geist der Quantentheorie*". What is this spirit? It is an expression of Bohr's mind. He has an imagination of unsurpassed power and flexibility, combined with perfect intellectual taste. He is incomparably swift in apprehending the point and value of any new idea. His apprehending mind turns with an infinite swiftness and arranges ideas into changing theories as Shelley's mind threw words into an empyrean to soar together into shining phrases. The opulence and discontinuity of his psychology seem related to the nature of the ideas of quantum theory. Is the possession of a quantum psychology by the creator of the quantum theory of the atom accidental? Perhaps the utmost command over discontinuous concepts is possessed only by those with a naturally discontinuous mode of thought. Continuous concepts may be used, similarly, only by those with a naturally continuous mode of thought. How far is a man's contribution to knowledge governed by the accidental relation between the type of his mode of thought, and the type of the ideas which happen to be developing in the intellectual environment of his day? Would Bohr's psychology have found as much scope during the reign of continuity in 1863 as it found during the reign of discontinuity in 1913?

Now that the Institute of Theoretical Research is more than a decade old its former students are numerous and eminent. At Easter many of them revisit Copenhagen to join the private conference on problems of atomic physics which is becoming a famous institution. Bohr's gifts of personality and leadership are seen fully on these occasions. Copenhagen is just sufficiently withdrawn from the centre of Europe to prevent any except enthusiasts from joining the conference. The number of attenders is perhaps thirty, but all are keen. Many of the best minds from Germany, France, England, Holland, Russia, are there. The orderly, quiet greyness of Copenhagen is not unsuitable to natural philosophical discussion. The Institute of Theoretical Physics is of a comfortable size, decorated in pale grey. Its lecture theatre is not too large, so that the discussions are saved from formalization. The programme is conveniently arranged in morning and afternoon sessions of one and a half hours. (So many scientific discussions are made wearisome by excessive length. Short and lively papers sustain interest.) Tea is served in the small library, on the tables at which famous books have been written and on which many games of ping-pong have been played during the relaxations of genius.

How can the appearance of the master be described when he arrives unobserved during a ping-pong battle, in order to announce the beginning of a fresh discussion in the theatre? He stands modestly at the open door, as if shaping to speak, but unable to find the correct expression. Within a minute or two his presence is observed, the game ends, and the players and audience file to the theatre.

In the theatre various experts report on the state of their subjects and their own researches. Heisenberg sits in the middle of the front auditorial bench and Bohr at the right-hand end, facing the speaker. Points of difficulty in the exposition will be raised by Heisenberg, who sits in isolation as the chief of the general staff, and others, leading often to enthusiastic interim discussions. During the lecture and these interim discussions Bohr will continually be rising and making comments and suggestions, in his quiet tones, and often no one will notice him for quite long periods. The leader is allowed to commune aloud with himself, without regard from his followers! Was there ever such a leader with so few of the ordinary qualities of leadership? The problems of quantum theory seem to fall under Heisenberg's command as soldiers under that of a general, while Bohr seems almost designedly to be causing disorder and confusion in the theoretical ranks. The effect is due to his imaginative subtlety. His comments often appear to be irrelevant to the point under discussion, and reveal the discontinuous quantum-like processes of his modes of thought. The third master mind in these conferences is that of Dirac. He speaks little, and then as a sphinx. His comments often contain an uncompromising dry humour, for there is quiet when he wishes to speak, and an oracular remark may come, followed by a puzzled silence among intellectually intimidated hearers, for Dirac speaks as though he were in communication with another world where the principles of physics are clearer than they are in this. His comments sometimes seem to cause the struggles of others with the principles of physics to appear as the flounderings of persons in the surf, learning to swim, must appear to a swimmer standing on the shore.

F. Bloch will speak brilliantly in several languages on several branches of theoretical physics. Sometimes he will be requested to explain other people's researches, as he can expound them better than the discoverers. And then the conference may end with an entertainment. On one occasion a parody of Goethe's "Faust" was produced. The co-operation of the artistic talents of the physicists gave a production which will never be forgotten by the audience. Goethe's songs were parodied in elegies on the con-

tradictions of the quantum theory.   Frau Kronig sang
these beautifully to the accompaniment played by
Heisenberg.   As each new contradiction was enum-
erated she introduced a sob into her voice.   Enter-
taining performances were given by Kramers and
others.   He expressed with infinite feeling an amus-
ing parody of the wrestlings in Faust's soul.

The Copenhagen spirit of the quantum theory
contains culture besides physics.   Niels Bohr grew
in a distinguished family in a most civilized city.
His father was professor of physiology in Copenhagen University, and his mother was a member of
the distinguished international family of the Adlers.
His brother Harald Bohr is professor of mathematics
at Copenhagen and a mathematician of international
reputation.   Bohr has said that his interest in physics
started at school, largely under his father's influence.
At the University he was guided by Professor C.
Christensen, of whose originality and endowments he
speaks very highly.   His first considerable piece of
research was a theoretical and experimental investigation of the surface tension of water by means of
oscillating jets.   Lord Rayleigh had explained that
the vibrations of a jet of water might be particularly
suitable as data for a surface-tension determination
because the surface of the jet of water is continually
being renewed while the liquid squirts forward, thus
removing the disturbing factor of contamination of
the surface, which destroys the accuracy of methods
depending on surfaces of static liquid.   Bohr refined
Rayleigh's and Pedersen's calculations and made the
experimental investigations in his father's laboratory.
This work was inspired by the offer of a prize by the
Academy of Sciences in Copenhagen, and was the
successful essay.   It was published in the *Transactions
of the Royal Society of London* in 1908.   Born in 1885,

Bohr was then twenty-three years old, and had demonstrated exceptional theoretical and experimental ability in a research typical of Lord Rayleigh. In recent years the already great reputation of Lord Rayleigh has steadily increased, which is interesting in comparison with the fate of the reputations of some of his great Victorian contemporaries. The twentieth century is tending to ascribe to him the position the Victorians ascribed to Lord Kelvin. When the discovery of the neutron was described in Chapter I, it was explained that Bohr deduced the rarity of interaction between electrons and neutrons by formulae analogous to Lord Rayleigh's formulae for the degree of scattering of light. After his essay in Rayleian physics Bohr became interested in the explanation of the properties of metals by electron theory. In the autumn of 1911 he went to England and studied the experimental researches progressing in the Cavendish Laboratory at Cambridge, and at the same time pursued his own theoretical researches. In the spring of 1912 he worked in Rutherford's laboratory at Manchester and afterwards became a lecturer in the University. He arrived during one of the most brilliant periods in scientific history. Rutherford had just demonstrated the existence of the atomic nucleus, which contained the main mass of the atom and was concentrated within a region extremely small compared with the size of the whole atom. This discovery was of immense consequence because it simplified the conception of atom structure extremely. As a first approximation the atom need not be conceived as a complicated, vague three-dimensional solid, but a structure built of points spaced apart. Compared with the atom, electrons and nuclei could without serious inaccuracy be conceived as points in space. Thus Rutherford's nuclear model was a tremendous aid to the visual imagination. How were these systems of points named atoms articulated so that they exhibited the well-known chemical and physical properties of atoms? Bohr felt that Rutherford's model of the atom had set the task reminiscent of the problem of the ancient philosophers: to reduce the interpretation of the laws of nature to the consideration of pure numbers. The prospect seemed thrilling with promise. But as Bohr has described in his *Faraday Lecture*, the first consideration of the possibilities of this model of atomic points showed that it could not operate according to known physical laws. It could not have the stability so characteristic of atoms, if it was articulated according to the laws of mechanics which govern the movements of familiar objects. The chemical properties of atoms of hydrogen and other elements are extraordinarily stable. Hydrogen from different parts of the earth exhibits precisely the same chemical properties. Hence gradual changes in the condition of atoms were improbable. The structure of an atom could not be slowly modified as the structure of the solar system is slowly modified by the impact of meteors and the perturbations caused by distant stars. If the electrons and protons in atoms behaved as the planets and sun behave in the solar system, each atom of hydrogen in the universe would be in a different stage of modification, and the constancy of the chemical properties of hydrogen would be inexplicable. Nor can the nature of the light emitted by disturbed atoms be explained by a gradualist theory of change in the relative positions of those electronic and protonic points constituting atoms. Atoms are very limited in their range of optical expression, they emit only light of a few definite wave-lengths or colours. This is incompatible with the vibration of any system of particles controlled by the laws of the movements of familiar bodies. Such systems ought to emit light of every colour, according to their state of modification, or disturbance. If a hydrogen atom, consisting of one proton and one electron, were operating according to laws similar to those governing the mutual rotations of the earth and moon, the proton and electron might be at any distance apart and hence vibrating in any degree and consequently emitting light of any colour. The exceedingly strict limits of atomic behaviour show that very powerful selective principles are governing the movements of the constituent electrons and protons; there is a profound absence of infinite variability in state. If the constituent protons and electrons were always in varying relative movement, so that the size of atoms was varying continuously over a wide range, solidity would be inconceivable. Matter would lose one of its most material attributes —solidity. The commonest properties of matter indicated that atoms must be remarkably stable, and limited in their possibilities of change.

Evidently some entirely new interpretation of the internal behaviour of atoms was necessary. The internal movements of an atom must be operated according to laws different from those of the movements of familiar bodies. Bohr's imaginative genius inspired him to see that the conceptions of the mechanics of the emission of light and radiation which had been forced on to Planck by his studies in the theory of the experimental facts of radiation might be applied in the explanation of the movements of the protons and electrons in atoms. He wanted to be able to assume that the changes in atoms are restricted in variety. This would follow if changes within the atom could occur only in finite amounts, if the amount of action within an atom on any occasion was restricted to a definite quantity, and not to one of an infinite range of quantities. Bohr saw that Planck's idea of quanta of action would serve as the controlling principle. He conceived the postulate, therefore, that any well-defined change of state of an atom is to be considered as an elementary process, consisting in a complete transition of the atom from one state to another. The transition was sudden. At one moment the atom was in one state, and at the next in another. While in any one of these states the atom remained absolutely unchanged, stationary. The stability of the materials of nature depended on the stability of the stationary states, and the restricted possibilities of transitions between these states, owing to the transitions having to operate in the limited currency of small multiples of a finite quantity of action.

The most astonishing aspect of natural phenomena is their simplicity. Why should the phenomena of a vast universe be intelligible, and why should any phenomena ever be repeated? Why should the principles of the universe not change with the passage of time? The remarkable regularity of properties among the infinite millions of material objects in the universe is an extraordinary fact, and indicates that limitation of possibilities is a fundamental aspect of natural phenomena. The idea of atomicity pervades the possibilities of material change besides the possibilities of material existence. The atom or quantum of action is more fundamental even than the ultimate atoms of material.

By relating the stability and limitedness of atoms to the existence of quanta of change Bohr made a great contribution to the formation of the new conceptions of the material universe. He explained the material philosophical implications of the existence of quanta of action, whereas his predecessors had used the idea as an ingenious key to physical problems, without incorporating it into their conceptions of the nature of matter and change. Considered as natural philosophers rather than physicists, Planck and Einstein are less original than Bohr.

The importance of Bohr's first postulate was philosophical; with it he accomplished the most difficult of cultural tasks, the changing of the general conception of natural phenomena. He was the chief converter of philosophers from the continuous to the discontinuous or quantum conception of the material universe. Planck's chief greatness is as a physicist. He discovered the existence of the quantum of action and to-day (1932) he is universally recognized as the senior theoretical physicist. But he did not perform the natural philosophical task of establishing the quantum conception of nature as the basis of the contemporary conception of the material universe. He was, as it were, a quantist in spite of his general conception of the material universe, whereas Bohr is a quantist because the quantum conception is in his processes of thought besides the external world. It is his natural way of interpreting phenomena. If he had flourished in 1863 he would probably not have been more than an eminent physicist, for at that time physicists had not toiled through to the quantum, and the natural quantum-thinker could not have found scope for the exercise of his highest genius.

Nevertheless, if Bohr's greatest achievement is found in natural philosophy rather than in physics, his achievements in physics are brilliant. This is exemplified in the practical consequences of his formulation of the second quantum postulate. According to this, the radiation or light emitted or absorbed, by an atom, which is the method by which the energy of an atom is decreased or increased, is always of a definite wave-length or colour, and this wave-length or colour is simply related to the quantity of energy released or absorbed.   This simple formulation enabled the almost indescribable complications of spectroscopy to be interpreted with swift success.   The achievement resembled the discovery of the clue to the Egyptian hieroglyphs.   Decades of spectroscopic research had contributed little more than an increase in data.   The wave-lengths of the radiations emitted by various atoms under various conditions had been exactly determined for hundreds of thousands of cases.   This information had an immense diagnostic value in spectrum analysis, enabling innumerable substances and atoms under a variety of conditions in various parts of the universe to be identified.   But the explanation of why they exhibited their characteristic labels of coloured light remained obscure.   Bohr's quantum theory of the behaviour of atoms was magnificently fruitful, and the harvest was gathered rapidly.   Within a few years his followers had solved those aspects of atomic behaviour to which his conceptions were applicable, and had arrived at difficulties which indicated limitations in these conceptions.   The Bohr atom had still many visual qualities.   Though the observer could not conceive how its state could change suddenly, so that an electron in one place at one moment could appear in another distant place at the next, he could conceive the atom in any of its stationary states.   These resembled a system of mutually revolving bodies such as the system of the sun and planets.   The Bohr atom could in part be visualized, though the mechanism of its changes could not.

Within the twelve years following the introduction of Bohr's model in 1913, innumerable refinements and alterations were incorporated, until the quantum model of the atom became a remarkable construction to which every interested worker added his own little collection of bolts and nuts and rods so that it could perform the particular motions he desired.   Another natural revision of conceptions was required, similar in spirit to Bohr's introduction of the postulates, but more radical.   This was supplied by Heisenberg. He devised a calculus for describing the observable properties of atoms without assuming any internal mechanical structure which could be visualized. Dirac worked on similar lines with perhaps even greater ruthlessness.   His natural psychology was adapted to such a point of view, and the scientific environment of 1925 was perfectly suited to the operations of his most powerful intellect.

In 1924 de Broglie had made a brilliant physical contribution to the solution of the difficulties offered by the Bohr atom.   He had suggested that the electrons and other particles might have a double property, and have the power of simultaneously resembling in some aspects waves, and in others particles.   His suggestion was ingenious and has the sort of place in the history of science occupied by Planck's suggestion of the existence of a quantum of action.   It was a device of the highest physical genius but was not closely related with a general natural philosophical conception of phenomena.   The resolute attempts to conceive the material universe in terms of the deepest principles of nature come from Bohr, his superbly gifted colleagues Heisenberg and Dirac and the other brilliant members of his school.

Following his resolution to base the description of phenomena on what is observable, without assuming any sort of unobserved structure in the object, Heisenberg arrived at his famous principle of *uncertainty*.   He explained that if atomic changes could occur only in atomic amounts the state of ultimate particles could not accurately be measured.   Changes in the motion or position of an electron could be measured only by the reception of rays of light or equivalent quantities of energy in an observing apparatus.   The energy of the ray of light would be of the same order as that of the observed electron.   Hence the electron would be disturbed by the ray of light used to illuminate it and its undisturbed state would remain not exactly observable.   A stone on a distant mountain side may be revealed by the beam of a searchlight because the energy of the beam measured by the pressure it exerts on the stone is negligible.   The stone is not pushed out of its place by the incident beam.   Within the atomic landscape the condition is different.   Any ray of investigatory light gives any obstructing electron a considerably disturbing push.   Accurate atomic-searchlight beams are impossible.   Since concepts such as space and time are derived from experience, atomic beings could not have derived the same concepts of space and time as are used in describing the behaviour of objects such as billiard balls and planets.   The latter are negligibly affected by the rays of light used in observing them, whereas atomic particles are not.   The concepts of space and time deducible from atomic phenomena are different from those deducible from the large-scale phenomena of everyday life.   An atomic being who had derived his conceptions of space and time from his atomic environment would find them as inapplicable to the description of the movements of, to him, immense objects such as a billiard ball, as the human finds his notion of space and time derived from experience of common objects to the behaviour of single electrons and protons.   The incomprehensibility of quantum conceptions is due to the provisional character of the traditional conceptions of space and time. In order to understand the universe, the existence of quantum properties, that is, the atomic nature of all natural phenomena including matter, must be accepted as given. They cannot be explained, as the existence of redness cannot be explained. When they are accepted the universe becomes more intelligible than ever before. In the opinion of Bohr a radical acceptance of the quantum conception of phenomena represents one further advance towards the objectivization of physical knowledge. He considers the philosophical value of Einstein's theory of relativity is in the purification of unconscious subjective elements from the Galilean and Newtonian conceptions of mechanical laws. Their natural philosophical achievements were of the same sort. It is foolish to ridicule Galileo because Einstein showed that subjective elements still remained in his law of motion. It is equally foolish to ridicule Aristotle because his law of motion contained even more subjective residues than Galileo's. Aristotle had to work with the words and ideas of his time and he showed himself a genius of extraordinary insight in producing conceptions containing less subjective assumption than those contemporarily current. Bohr regards the attempts on the basis of his work to read more subjectivity into the interpretation of natural phenomena as a misrepresentation of his views. In his opinion the clearer recognition of the rôle of subjectivity in physical observation is an objective philosophical discovery of great importance. The work of his school is to introduce ever more objectivity into the conception of phenomena. He considers, therefore, that it is entirely opposed to mysticism. The tendency to suppose that people may believe anything they like because traditional space and time are not exactly applicable to electrons exhibits a complete failure to understand what are the chief contributions of his own school of thought. The human intellect requires a consistent, coherent description of the universe, and the quantum theory satisfies this requirement more beautifully, consistently, and completely than any other theory yet advanced. Bohr does not sympathize with Planck's opinion that the contradictions of the quantum theory are apparent and due only to the limitations of the human mind and would not appear contradictory to a super-mind. Nor can he sympathize with Einstein's attempts to show that quantum phenomena can be interpreted as special cases of classical phenomena. He believes such attempts show that the scientist believes in quanta only as devices and not as fundamental mechanisms of the physical universe; they show, as it were, the failure to feel quantically in spite of being able to think quantically. The quantum theory has exposed unexpected subjective elements in the traditional conceptions of space and time, and has therefore introduced more objectivity into science. Bohr considers we shall always think of the world and analyse it in terms of classical mechanics because classical mechanics is a refinement of the common-sense description of phenomena. The modes of thought natural to classical mechanics are developments of the modes of thought produced by the ideological evolution of an animal of a certain size adapted to live in a world of objects of a certain size. The human mind cannot visualize quanta because it was not evolved in an environment of individual quanta, it has evolved in a world of objects such as flies, trees, stones, of an order of size similar to that of its own associated body.

The notion of exact measurement has evolved in an environment where the interference by the means of measurement with the object measured may often be negligible. Classical mechanics, classical notions of space and time, will always be the human mode of exact description. They are the only technique for giving unambiguous answers to scientific questions. But in order to speak unambiguously we must pay a price, we must ignore the excessively minute. If we wish to speak with full accuracy, then we must always be slightly ambiguous.

It is of interest to note that Bohr does not see any necessary connection between the ambiguity of quantum phenomena and human free-will. He believes the problem of life may be beyond us. The machine is dead because we understand it; we should appear to ourselves dead if we understood ourselves, but in the last analysis we cannot know unambiguously how every atom in the body behaves, because of quantum limitations. He comments that too often the most difficult ideas in science are popularized, instead of the easiest. If only the difficulties in the ideas of mass, relative motion and quanta were expounded, so that the general reader could appreciate the obscurities in these common notions, how much more modest we would be in assertions concerning more difficult problems. Bohr remarks that Heisenberg's principle of uncertainty does not imply a one-sided departure from the ideal of causality underlying any account of natural phenomena. In fact, the persistence of the law of conservation of energy through quantum changes is a most striking example of the persistence of causality, especially when it is remembered that the concept of kinetic energy was deduced from a motions-in-space-time picture which is inapplicable to quantum phenomena.

The sort of ambiguity in the possibility of the description of quantum phenomena is not at all new in human thought. It is seen, for instance, in the perception of a musical harmony. The length of the harmony and its comprehension are mutually exclusive below a certain limit of shortness, for if the notes are too few the harmony cannot exist. It is seen, also, in the classical definitions of temperature. This depends on the average condition of a large number of particles. The precision of the definition becomes less and less as the number of particles considered decreases, until it has no meaning when applied to a single particle. As the definition of temperature contains an element of ambiguity the reversal of the general levelling of temperatures in the universe is not impossible. The drift of the material of the universe towards one uniform temperature is not an absolute necessity, arising from the nature of things. The tendency of temperatures to uniformity does not imply that the universe is absolutely necessarily running down, but that the prediction of any reversal, a tendency for the universe to be wound up, cannot be made from a study of the temperatures of the various bodies in the universe. The reversal would have to be due to events not of a thermal nature.

The physics laboratory at Manchester University must have been a most extraordinary place in the year 1914. The chief creative experimental atomic physicist and the creative theoretical atomic physicist were both working there, and half a dozen others who would have been outstanding in almost any other environment. In 1916 the Danes recognized the genius of their greatest living man and created a chair for him in Copenhagen University. In 1920 the Institute of Theoretical Physics was founded for

him by Danish subscription and extended by gifts from
the Rockefeller Foundation. In Denmark science
receives much aid from the Carlsberg Foundation.
The profits from the famous Carlsberg lager breweries
are devoted to the encouragement of science and
culture. Besides providing buildings for laboratories,
research grants and other financial aids, the Carlsberg Foundation provides a house of honour for the
leading Danish savant of the day. Bohr has recently
been installed in this house as the successor of the
philosopher Höffding, at the early age of forty-seven
years. May his spirit continue to lead and inspire
natural philosophy!

CHAPTER III

SOVIET PHYSICS

I

EXCELLENT physical research is being done in Soviet Russia. The theoretical explanation of radioactivity given by Gamow of Leningrad has already been mentioned. Mandelstamm independently discovered some aspects of the phenomenon known as the Raman effect. Indeed, he was aware of Smekal's deduction of the existence of such a phenomenon from quantum-theory considerations and was consciously looking for it. Raman discovered the effect by pure experimental observation, and great credit is due to him for his energetic extension of investigation into the phenomena and the impression of its importance on the scientific world. The physical investigation of the nature of the much-discussed mitogenetic rays has been vigorously pursued in Soviet laboratories. The existence of rays emitted by the cells of certain animals and plants when the cells divide, or mitote, in the progress of growth, into two new cells, was asserted by Gurvitsch. He believed he had shown that when cells in a root of an onion divide, they emit rays which can stimulate the rate of division of the cells in a neighbouring onion root. He considered, also, that these rays must resemble those of ultra-violet light, as they could be intercepted by a sheet of glass but penetrated a sheet of quartz. For some years Gurvitsch's results were received with general scepticism. He had detected the rays by their biological effects, by the stimulation of growth and cell-division in other organisms such as plant roots and yeasts. Physicists were sceptical of the existence of physical agencies, such as rays, which could not be detected by physical methods. The recent astonishing improvements in the sensitiveness of physical apparatus suggested, however, that a special physical apparatus capable of detecting the mitogenetic rays, if they really existed, might now be constructed. Professor Joffe and other Soviet physicists attacked the problem of designing such an apparatus. They designed a special form of Geiger's counter. This apparatus registers the appearance within it of a single electron, and counts the sequence of electrons automatically. The sensitivity of a Geiger counter may be illustrated by comparison with the sensitivity of the most delicate chemical methods of analysis, in which quantities containing less than a million million atoms are undetectable. The Geiger counter offered to the physicist an apparatus far more sensitive than any chemical apparatus and possibly sensitive enough to equal the sensitivity of the onion roots and yeasts Gurvitsch had used to detect the mitotic rays. As even the most delicate biological changes involve large numbers of molecules the Geiger counter, which could detect the presence of an electron, which is much smaller than a molecule, might reasonably be expected to reveal the rays, if they existed. That the mitotic rays might be strong enough to affect living organisms but too weak to affect the older forms of physical recording apparatus is not surprising. The potency of exceedingly small quantities of chemical substances such as vitamins, hormones and ferments in causing profound changes in living organisms shows that biological tests for their existence might be more sensitive than any technique of testing which required for successful operation groups containing more than a few molecules. Gurvitsch had made spectroscopic examination of the rays by a biological method. By directing the rays from a growing yeast culture through a spectroscope he was able to determine their wave-length. The optical system was made of quartz and the quartz revolving prism refracted the rays through an angle appropriate to their wave-length or colour. Their presence at the observing end of the spectroscope was shown by placing a yeast preparation instead of the eye or a photographic plate in front of the eye-piece, and counting the rate of cell-division. If there was a well-defined increase for any particular angle of the spectroscopic telescope the existence of a band of mitotic rays of appropriate wave-length could be demonstrated. The scientific world found Gurvitsch's technique of placing yeasts instead of photographic plates in front of the registering end of a spectroscope rather bizarre. Many attempts to repeat his experiments failed. Gurvitsch considered his measurements showed that the rays were of the ultra-violet and of about half the wave-length of visible light.

The Soviet physicists commenced their investigation by studying the biological properties of ultra-violet rays of similar wave-length from a quartz mercury lamp. They gradually decreased the intensity of the rays from the lamp until it had been reduced to about a thousand millionth of its initial value, until it was too slight to be able to affect a photographic plate. Yet the biological effects of ultra-violet rays from a lamp through this range of intensities remained much the same, and quite different from those which Gurvitsch had ascribed as due to mitotic rays. At this stage they introduced the use of a Geiger counter, which had also been done independently by Rajevsky at Frankfurt. The Geiger counter, because it can register single particle effects, is far more sensitive than a photographic plate. It was made of an aluminium plate and a platinum wire fixed one twenty-fifth of an inch away, suspended in a vacuum vessel with a quartz window. When ultra-violet rays pass through the window and fall on the aluminium plate they cause electrons to be ejected from it. If a strong electrical field of about 400 volts is applied to the gap between the aluminium plate and the platinum wire any ejected electrons within the region are caused to shoot forward with increased energy. As the accelerated electron shoots forward it loosens a large number of electrons from the atoms of the surrounding rarefied air, so that the chamber accumulates a relatively large electric charge. This may be arranged to operate a radio loud-speaker, so that each electron ejected from the aluminium plate by the ultra-violet rays will cause a click in the loud-speaker, and the frequency of clicks will measure the rate of ejection of electrons and hence the strength of the ultra-violet rays impinging on the aluminium plate.

The existence of ultra-violet rays emitted by living cells could be investigated by placing the tissue containing the cells in front of the quartz window. When this was done with cells from the roots of onion, yeast and frog's heart the frequency of clicks in the loud-speaker rose very definitely by 20 to 70 per cent. When the muscles of a frog were caused to contract by electrical stimulation the frequency of the clicks increased by 230 per cent for the strong muscles of autumn frogs and by 60 per cent for the exhausted cells of spring frogs. A determination of the wave-length of the rays emitted by muscles confirmed

Gurvitsch's results.   None of these emissions of ultra-violet rays were detectable by photographic plates.

The Soviet investigators made an excellent extension of their research into the emission of rays during chemical reactions.   They avoided the tendency to regard the research as exclusively biological, or concerned merely with the physical properties of biological organisms.   This was an admirable feature of their work.   Professor Joffe was concerned to explain that this research had an important methodological interest. He has commented that the relations between biology and physics are not yet clearly defined.   The electrical currents associated with the passage of messages along nerves, and the electrical effects caused by muscular contradictions such as those of the heart, have been studied fruitfully, and the use of the apparatus for measuring currents caused by the heart, named the electro-cardiograph, has had very important applications in the clinical treatment of heart disease. The value of ultra-violet ray treatment of the skin in consumption and other diseases is now a commonplace, yet the way in which the rays affect the cells of the stimulated body remains obscure.   The way in which the muscles and nerves produce their important and instructive electrical currents is equally obscure.   The explanation of the intimate mechanism of reaction between physical agents and biological cells now requires to be explored.   Until it has been successfully explained the relations between physics and biology will remain distant and disjointed.   Professor Joffe considers the refined study of the interaction between ultra-violet rays and cells in process of growth or activity is a useful contribution towards the methodological aim of integrating physics and biology, and the destruction of traditional prejudices which tend to limit the field of physical research to the properties of the traditional objects of physical research, such, for example, as the properties of metals or crystals, and the restriction of biological research to the traditional biological aspects of living organisms, such as their shapes, evolutionary history and reproductive systems. Free from the prejudice of the physical researcher to restrict his attention to physical objects, and of the biologist to living organisms, the Soviet investigators had no difficulty in overcoming the mental prejudice against investigating purely chemical reactions in order to discover whether they might emit rays similar to those emitted by dividing cells. They found many chemical reactions were accompanied by the emission of similar feeble ultraviolet rays.

Professor Joffe considers the Gurvitsch mitotic rays are ultra-violet rays of wave-length between 2,000 and 2,400 units. (The wave-length of visible light is about 5,000 units.) They are produced during the chemical reactions in the cell concerned with processes of oxidation and glucose reactions. Their intensity is about one million-millionth (one-billionth) that of the ultra-violet rays from an ordinary mercury lamp. These rays influence the life of cells in many ways, accelerating growth, and perhaps causing genetic mutations leading to new species. He believes the researches will be valuable also in reducing the attractions of vitalist philosophy which flourishes on the obscurities of the intimate nature of the processes of living organisms, and also will help to destroy inadequate mechanical theories of phenomena peculiar to living bodies.

The Soviet mitotic ray researches have not yet received universal acceptance, but in his Annual Address for 1932, to the Royal Society of London, the President, the famous biochemist and discoverer of vitamins, Sir Frederick Hopkins, said research by many during the previous year had suggested that chemical reactions in living tissues are indeed accompanied by radiations. He considered that the phenomena may be related to the luminescence noticed in many non-biological chemical reactions and may have affinities with the phenomenon of luminiferous organs in certain animals, and in luminiferous bacteria. He quoted the claim that the characteristic spectrum of radiation emitted from a tetanized or stimulated muscle is identical with that yielded by the chemical breakdown of creatine phosphate, which is known to occur in active muscle. He considered the existence of rays emitted by certain living cells as proved, though he was unable to say whether the phenomenon is of fundamental importance or even whether such rays are associated with all forms of life.

II

Particular researches such as the discovery of new rays and new theoretical explanations of phenomena are not in themselves different in sort from physical researches conducted in research departments outside the Soviet Union. The social conditions under which the researches are done are, however, unique. The organization and atmosphere of Soviet institutes is often distinctly different from that of institutes in countries whose inhabitants live within a different type of social order, and this aspect of Soviet scientific work is perhaps the most original. It is exemplified well in the fine Physico-Technical Institute in Kharkov (see Plate V, p. 57). This Institute was opened in 1930, one of a group of physical research institutes included in the scientific sections of the famous comprehensive plans for the development of the Soviet Union. A decade ago there were only two important physical research institutes in the Union. These were quite inadequate to the training of the large number of physicists required for the dissemination of scientific culture through a population in course of cultural revolution. Before many new institutes could usefully be founded the training of a corps of future directors was necessary. This was accomplished by building a very large physical institute where ability could be concentrated and trained. This was the Physico-Technical Institute at Leningrad, directed by Joffe. Within a few years 200 research physicists had been trained for useful work. This served as a nucleus for the staffs of the new institutes to be built in other provinces from which the cultural spirit of research could be disseminated. For instance, the Physico-Technical Institute at Kharkov could serve as the centre for the province of the Ukraine. The building of the Institute at Kharkov had much more than a local influence. The presence of qualified colleagues enabled the foundation of a new laboratory at Dnieperpetrovsk to be successful. If the Kharkov institute had not already been founded physicists would not easily have been persuaded to go to a virgin cultural area such as Dnieperpetrovsk. A scientist usually requires the stimulation of conversation and discussion with colleagues in the same branch of research. He is often discouraged if he lives in a new institute in a new city where no one understands his aims or difficulties. As Dnieperpetrovsk is only six hours' journey from Kharkov (distances in the U.S.S.R. are so vast that a Russian regards a six hours' train journey as an Englishman would regard a one hour's train journey), physicists could be persuaded to staff the new Physico-Technical Institute there, after the Kharkov institute had been started.

The Physico-Technical Institute at Kharkov contains between two and three hundred workers of all grades. About fifty of these are qualified professional physicists, and about twenty are research students. There is no systematic tuition, but there is a theoretical seminar. The Institute belongs to N.I.S., that is, the Scientific Research Sector of the People's Commissariat of Heavy Industry, lately directed by N. Bukharin. The dependent institute at Dnieper-petrovsk is about half the size of the Kharkov institute and is included in the group of three or four physical research institutes analogous in character which have been erected in different centres in the U.S.S.R. and attached to N.I.S. The association of the laboratories with heavy industry and not with the universities and academic institutions is interesting, and has certain effects on the atmosphere of the institute and the choice of research problems.

The director is Dr. Leipunsky, who is a Communist. Of the fifty professional scientific research workers in the laboratory, about six are Communists. Many of the large staff of mechanics in the workshops are Communists.

The Institute is divided into departments, and the workshops count as independent departments. This contrasts remarkably with practice in other countries, as outside the U.S.S.R. a laboratory workshop would not have as much independence as the departments with which they were connected. In the West the laboratory workshop is regarded as a servant rather than a colleague of the research-rooms. There are ten departments. The Institute is considered to be too large, and in future will probably be sub-divided. New physical institutes will be designed as smaller units. The size of the Kharkov institute is due to the initial policy of concentration in order to create a physics personnel. As the number of persons qualified to direct institutes increases, the tendency will be to build a larger number of smaller autonomous institutes, instead of a few large ones with many departments. The chief departments are for research on high voltages, low temperature, X-rays, magnetism, photo-electric cells, solid bodies, electro-technics, and physical theories. There is a department of administration and accounts, which is much more important in the life of the Institute than analogous departments in a Western institute.

The workshops are arranged in a group of three for woodwork, metalwork and glass-blowing. They are under the direction of a manager who ranks as a departmental director. Each shop has a master-worker or foreman with the exception of the glass, which has two. A physicist in any of the research-rooms who requires a special glass tube to be blown would proceed as follows. He would write an order addressed to the glass-blower's workshop, but take it to the general manager of the workshops. The general manager writes his own version of the order, estimates how long it will take to be done, signs it, and gives it to the particular workshop master. This system prevents the workshops from being disorganized by importunate individual research workers. It distributes the work equably, and generally works well. Soviet laboratory workshops are notable because of the social prestige of the manual worker. Public opinion regards his social class as higher than that of the intellectual research worker, though he has no superiority, in fact often inferiority, to the research worker in material living conditions. The shortage of scientific instrument factories in the U.S.S.R. makes a laboratory's own workshops much more important, as often the most difficult instrument manufacture must be done there, if at all. The relation between the research departments and the workshops in the Kharkov institute resembles on a smaller scale the relations between, say, an English laboratory such as the Royal Institution and an instrument manufacturing firm such as Adam Hilger Ltd. The highly qualified, and indeed scientifically distinguished, staff of Adam Hilger are called in as collaborators rather than servants in the design of apparatus. The manager of a Soviet laboratory workshop has something of the authority of the proprietor or managing director of a high-grade scientific instrument firm. The workshops contain sections for electrical wiring and fitting, plumbing, etc. The metal shop has a staff of about fourteen, and there are about ten glass-blowers.

The most original of all the features of a Soviet laboratory is the system of planning the course of future research. Each department is required to draw up a general plan for one year's work, from January 1st to December 31st. For each quarter the plan must be given in greater detail, and there must even be a plan for each day's work. A research worker must keep to the subject he has entered in his plan. If he wishes to change his subject, almost the whole of the institute must discuss his proposal before he receives permission. At regular intervals, which in practice prove to be of about one month's duration, the research worker must assess what percentage of his plan has been accomplished. This usually proves to be 80–90 per cent. The personal assessments are extraordinarily honest.

The planning of work causes a rhythm in performance. After three-quarters of a planned period has passed the work done is usually considerably behind the estimate. A heavy spurt is made in the last quarter so that a large percentage of the plan is nearly always completed, and sometimes surpassed. For instance, liquid helium was planned to be produced by January 1st, 1933. Actually liquid helium was produced one month earlier. This was done by the young German low-temperature physicists, Dr. Martin and Dr. Barbara Ruhemann. In 1932 the authorities of the institute decided to inquire for a foreign expert to assist in their liquid helium researches. Their choice fell upon Ruhemann, who at the time was working with Professor Ewald at Stuttgart. Previously he had worked in Professor Simon's low-temperature laboratory in the Bodenstein Institute in Berlin. Later Simon went to Breslau, and is now in Oxford. The Ruhemanns have found the living conditions and opportunities for work better at Kharkov than in the present depressed Germany. Besides operating their helium apparatus successfully, which they regard as nothing more than a routine, they have a large plan of research concerning the specific heats of gases and the spectra and magnetic properties of substances at low temperature.

The departments have contracts with the industrial trusts to procure scientific data of industrial value. The low-temperature department's industrial contracts amount to 100,000 roubles at the moment. This sum may be drawn upon for the purchase of equipment and general development of the institute.

The distribution of industrial research work is organized by N.I.S. The industrial trusts and factories send their problems to the office of N.I.S. in Moscow, where the suitable institutes for the various required researches are chosen.

The workers in each department or research room are organized in a brigade. This holds frequent meetings, in which the work of the brigade, the general problems of the institute, and almost every conceivable question, are discussed.   The needs of the brigade and the institute are earnestly debated, and immense efforts are made to help any member of the brigade who is specially advancing the accomplishment of the brigade's plan.   But the personal desires of a member of a research brigade receive little notice. Everything possible is done for the research worker realizing a communal aim, but very little for the worker realizing a personal aim.   There is remarkable subordination of the individual to the community. As an individual no one in the institute counts, yet individuals may receive much honour and compliment because they are realizing communal ends.   This is the explanation of the curious combination of personal fame and communal ideals noticeable in the U.S.S.R.

Every brigade has to give an account of its economic besides its scientific activity.   There is a cheque book associated with each research problem so that the cost of the investigation may be automatically recorded. The cheque book accompanies each order to start on a problem, and the actual cost is compared with the estimate in the brigade's or department's plan. The cost should never exceed the estimate.

Individual departments buy and sell from each other and then have a day of reckoning.

Workers have fixed salaries, but the cheques are made out by the hours of labour.   Thus a slow glass-blower will increase the cost of a research which involves his services, and his slowness will be apparent from the cheques.

This elaborate system of planning and recording requires a large administrative department, and the unusual importance of this side of the laboratory organization, compared with the analogous side in Western laboratories, explains why the administrative director is important. He ranks more or less as third general director, and is assisted by eight accountants, a cashier, and clerks. He superintends the book-keeping and the management of the stores. He also controls the important residential accommodation for the staff within the laboratory grounds. These include four large blocks of flats. He superintends the furnishing and running of these houses. The supply of heat, light, gas, water, electricity and repairs is one of his responsibilities, and he has a staff of handy men to assist in this work. In some ways he is to be compared with the bursar of a college at Oxford or Cambridge. The purchase of instruments and material from foreign countries is supervised by a special official. At present very little apparatus is purchased from abroad. An order for a few hundred pounds' worth of foreign apparatus is a serious matter, which usually must be discussed by several important committees in Moscow before being sanctioned. This is in contrast with the policy of a few years ago. When the laboratory was built large sums were spent on the purchase of foreign apparatus.

The departmental brigades are not of equal size. In this laboratory the low-temperature and high-voltage brigades are particularly large and are associated with a special workshop.

Most of this information was given by the kindness of Dr. Ruhemann. He finds the system of the laboratory interesting and enjoys working within it. His experience is that nearly everything that is required comes sooner or later, if the system is operated with patience and good-will. The informality of the personal life of the staff is attractive. Research workers may go into their laboratories at any time of the day or night. Some never start work before two o'clock in the afternoon and continue until the early morning hours.   Others are rarely seen in the laboratory in the evening.   As the flats are within the grounds, workers may walk to their homes or to the laboratory conveniently, and as often as they wish.

Many interesting researches are in progress.   Mme. Prichotjko is doing some beautiful work on the absorption spectra of thin films of solids.   Dr. Obreimov had thought of this work ten years ago, but at that time there was no suitable laboratory in the U.S.S.R. where it could be started.   The object of the research is to investigate the influence of the solid state on the spectra, especially of compound organic substances such as benzene, naphthalene and anthracene.   He is not so much interested in the systematics of the absorption lines, the exact measurement of their position, etc., as in the interpretation of their peculiar features. What is the meaning of the broken lines, the breadths of lines and fine structure?   The discovery of new ideas concerning the interpretation of the spectra of solid bodies, and the influence these might have on the conception of the structure of matter, may lead to the discovery of new techniques of research. Obreimov considers the discovery of new techniques of research is the chief task of the present generation of physical scientists.   He believes Kapitza, Semenov, Chernichev and other physicists and chemical physicists approaching the age of forty years will contribute much more in the discovery of new technical methods of research than in remarkable new fact.   Their inventions of method will probably be more impressive than the results they will themselves obtain through these new methods.   Their successors will collect the new orders of scientific fact.   By considering the peculiarities of the spectra of solids openly, and in a perspective different from that of present spectroscopy,

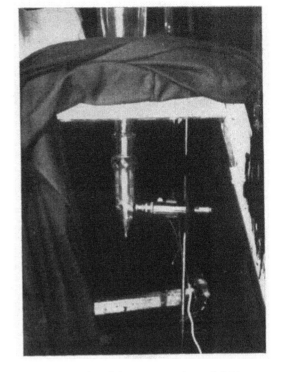

A rotating anti-cathode X-ray tube. Physico-Technical Institute, Kharkov. (See page 91.)

Apparatus for disintegrating atoms of lithium. Physico-Technical Institute, Kharkov. (See page 90.)

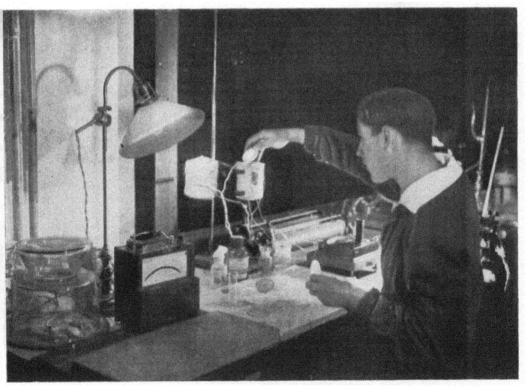

Preparing thin solid films for absorption spectral examination. Physico-Technical Institute, Kharkov. (See page 88.)

an advance may be made towards the discovery of new methods in spectroscopy. He considers freedom from traditional thought the most difficult condition of scientific discovery. When Cockcroft and Walton made their electrical disintegration of atoms they found the disintegrations commenced at a surprisingly low voltage, about one-fifth or one-tenth of the expected voltage. Landau of the theoretical department in the Kharkov institute had shown that disintegrations might be expected at surprisingly low voltages, but traditional consideration of thought prevented experimentalists from being inspired by these calculations to try experiments in conflict with their prejudices. Thus the experiments of Cockcroft and Walton had been easily within the possibility of experimenters in several laboratories in various parts of the world, but time had to pass before any of them found the boldness to act in conflict with the traditional prejudice that a particle travelling under an electrical pressure of at least 1,000,000 volts would be necessary to accomplish an atomic disintegration.

In order to obtain absorption spectra from anthracene and naphthalene extremely thin crystalline or single crystal layers are prepared, not more than a millionth of a centimetre thick, and invisible to the eye. The crystals sublime unless kept at extremely low temperatures. Anthracene layers will last several days, but naphthalene layers sublime within a few seconds. The naphthalene layers are kept in naphthalene vapour. The orientation of the crystal is determined by a combination of X-ray and optical measurements. The spectra are obtained with light from a continuous spectrum, with exposures of ten minutes to half an hour, the wave-length varying from 3,060 to 1,200 units. Different absorption bands are obtained according to the orientation of the crystal layer. The qualitative similarities and differences between the solid, liquid and gas spectras of hydrogen and organic compounds are the chief interest of the investigations, and the preparation of the thin films requires extreme manual dexterity, in which Mme. Prichotjko is gifted.

In the high voltage department experiments with many different forms of apparatus are being made. Cockcroft and Walton's experiment of disintegrating lithium atoms by swift protons has been repeated by Senelnikov and his colleagues. High-tension discharges of 900,000 volts and ten amperes, lasting one ten-millionth of a second, have been produced by suddenly switching parallel condensers into series. Swift protons have been obtained from a Lange tube, the protons being produced by the creeping current effect along the wall of the tube.

The theoretical department contains research workers such as Landau and Rosenkevitch. Besides his research Landau is also engaged on the reconstruction of the course in mathematical physics in the Kharkov Technological Institute. He is attempting to work out a method of presenting the appropriate matter from a modern point of view, and eliminating methods and subjects which remain in the course only through historical inertia. There is an opinion that the teaching of mathematical physics is in general far too traditional and repeats old theory merely from academic habit, though modern theory could give a simpler and livelier presentation.

Strelnikov has succeeded in improving his rotating water-cooled anti-cathode X-ray tube. This gives extremely intense beams of X-rays capable of penetrating a great thickness of steel and other dense materials. X-ray photographs can be taken very swiftly with its assistance, because the rays are so

intense. With a little more development the production of an X-ray cinema film will become possible.
This can be done if X-rays can be produced sufficiently
intense to take photographs in rapid sequence. A
film exhibiting the change in crystal structure of a
piece of metal or other material under stress would be
extremely interesting. The biological applications of
an X-ray cinema camera readily occur to the mind.
An X-ray film of the changes in the molecular structure of muscle during contraction might be very
instructive. Dr. Muller of the Royal Institution,
London, has worked on the difficult problem of the
design of a rotating anti-cathode X-ray tube. The
intensity of the beam of X-rays from an ordinary tube
is limited by the resistance of the anti-cathode to heat.
The X-rays are produced by the impact of cathode
rays or electrons on a metal object, named the anti-
cathode, and the continuous bombardment by the
stream of electrons of one spot on the metal anti-
cathode may soon raise it to melting point. By
rotating the anti-cathode a fresh spot of cool metal
may be continuously presented to the electron
stream, and no spot will be exposed long enough to
become over-heated. The difficulties in manufacturing
such a tube are of a mechanical character. The
rotating metal anti-cathode must run extremely
smoothly, so that the beam of X-rays from it always
travels in the same direction, though always proceeding from a different spot of the metal. The
preservation of a good vacuum in a tube containing
a large swiftly rotating heavy object offers considerable
engineering difficulties.

The most remarkable feature of this interesting
institute is the youth and enthusiasm of the research
staff. Of the fifty research workers not one is over
forty years of age, and only four or five are over thirty

years of age. The results of the collaboration of these young enthusiasts in operating new systems of organizing scientific research, of receiving opportunity and responsibility at an early age, should prove not only of general interest, but of importance for the insight they should give into the proper methods of organizing cultural civilization, for this is becoming one of the chief problems of modern life.

III

A new Institute of Mathematics and Mechanics was opened in Kharkov in 1931. It contains excellent lecture-rooms, theatre and library, and a mathematical instrument laboratory is to be equipped in the future. It is larger than any mathematical institute in England. The director is M. Orlov, a well-known writer on the approximate numerical solution of integral equations, aerodynamics and allied researches in mathematical physics. He is aged thirty-three, and a Communist.

The elaborate plans of mathematical research are a remarkable feature. There are half a dozen sectors, branches of study, including the algebra of real and complex functions, theory of functions, mathematical physics, mechanics, geometry and probability theory. Each sector has had a plan of work for 1932, and a plan for 1933 which will be part of a general plan as their contribution to the Second Five Years' Plan. The sector on real function theory is led by the eminent mathematician Bernstein. He is now aged fifty-two years, which is very senior in a Soviet institute. He has one fully qualified mathematical research colleague and five assistant research students, termed aspirants (aspirants for the equivalent of a doctor's degree). Bernstein's plan of research for 1932 included studies in the approximation to functions, and orthogonal

polynomials. Zinssov is another of their senior
lecturers. He is aged sixty-five. Orlov himself leads
the sector of mathematical physics. In their 1932
plan they included researches on the differential equations concerned with the dynamical stability of aero-
planes; and the theory of gunnery, ballistics. For
their second Five Years' Plan they have proposed
studies including non-linear differential and integral
equations, graphical methods for solving differential
and integral equations, Bessel functions and elliptical
integrals, and mathematical instruments. By November
1932, 75 per cent of their plan for 1932 had been
accomplished, and they hoped 100 per cent of their
plan would be completed by the end of the year. The
research leaders choose their own problems for investigation and incorporate them in their plan. In this
institute they specialize on the theory of the mathematics used in technical work, such as aeroplane
designing. The theoretical interest of their research
is high. They are anxious that problems of practical
mathematical interest but little theoretical mathematical interest should not dominate the institute's
work.

In 1932 they had nine senior lecturers, nine junior
lecturers and thirty-six aspirants. In 1931 the figures
were five, nine and eighteen respectively. Their
senior lecturers deliver courses also in the Kharkov
Technological Institute. Twelve qualified engineers
have been sent to the Mathematical Institute to engage
in research. Their salaries are paid by the industrial
trusts that employ them. In this way the institute
receives 20,000 roubles annually from industry. The
total cost of running the institute is 150,000 roubles
(£15,000 at par) per annum. Some of their aspirants
work half-time in a factory, where they may be on the
engineering staff, and half-time in the institute. They

have an electrical engineer who works in an electrical factory in the morning and in the institute in the afternoon.

Besides M. Orlov there are nine other Communists on the staff. Eleven of the aspirants are Young Communists, that is, they are members of the junior organization of the Communist Party for persons under the age of twenty-five.

In some instances an industrial trust will send one of its staff to investigate in the institute a problem in which it is interested.

IV

Investigation of the properties and manufacture of refractory materials involves much application of physics. The new Silicate Institute in Kharkov is designed for the study of clay, earthenware, blast-furnace bricks, cement, etc. The development of heavy industry such as iron smelting has created a demand for furnace linings, etc., and hence a demand for the highly-developed technique of manufacturing these products. The convenient raw materials in the U.S.S.R. for the manufacture of refractories are not always the same as those used in America and other countries. Research is necessary to determine how the Soviet materials may be made into products as good or better than the conventional refractory products. The buildings and equipment of the Silicate Institute are first-rate, and might impress many observers as magnificent. As in most new buildings in the U.S.S.R., the corridors and rooms are generously proportioned. (The use of the metre instead of the yard as the unit of length is always allowing the Soviet architect three inches more than the British yard. The cumulative effect of this extra amount may be considerable. There is a tendency also to use half a metre where an English architect would use a foot.)

The walls are nearly always washed in dull white, with a roughish surface, and the paintwork is often pale grey. The windows are double-glazed to preserve warmth during the Russian winter, and all buildings are warmly heated. The proportions, restrained colouring and warmth often give a pleasing effect.

About 75 per cent of the institute's work is devoted to research. The staff includes about 200 persons, of whom eighty have a scientific qualification. There are two general sections, the first for refractory materials and acid-resisting materials; and the second for cement.

An inspection of the departments showed they were provided with equipment of the latest design. For example, there was an excellent equipment of electric furnaces. Considerable use was being made of furnaces of original design burning crude oil. The firing arrangement seemed very simple, as it consisted merely of a stream of drops of oil running into the bottom of the furnace, where they fired spontaneously. There were several laboratories containing examples of the various types of grinding machine for producing powdered materials. There was a special room for determining the size of clay particles by several methods, such as the rate of fall in liquids; electric furnaces whose temperature is regulated automatically by mercury cut-out governors, an excellent geological section for the examination and exhibition of raw materials, and large chemical laboratories for general analysis. There was a good library, with an English-speaking librarian, and the leading American and British periodicals were taken. The institute had a large room usable as a meeting hall, and in this had an exhibition of illuminated transparencies (photographic positives) showing the institute's equipment and work. In addition to the research laboratories the institute was being provided with a large factory-laboratory, where processes could be tried on a full industrial scale.

Among the products shown as results of research were graphite bricks for use in open-hearth steel smelting furnaces, heat-resisting bricks containing 70 per cent of alumina, contrasting with the normal alumina content of 40 per cent, and coke briquettes for use in blast-furnaces.

If the scientific talent of the staff is equal to the opportunities for research, much good work should be done in this fine institute.

CHAPTER IV

THE STARS AND THE UNIVERSE

I

THE remarkable expository gifts of Sir Arthur Eddington and Sir James Jeans have secured public notice for many of the interesting theories prompted by recent studies in astronomy and physics. Distinguished scientists even when they possess expository talent do not often seriously address themselves to the task of exposition until their first period of creative work is finished. After twenty years of intense research they may rest a little and wonder at what they and others have discovered. The impulse to wonder and to express beautifully the knowledge they have learned is not allowed to act during the chief period of creative research. No distractions are permissible then, when the researcher is discovering new things about the universe and experiencing the keenest of emotions. A distinguished mathematician has compared the appetite for the prosecution of research with that of the tiger for blood: once new knowledge has been tasted, never before known to man, the attractions of other things seem insipid, and the researcher seeks to tear further knowledge from the mass of obscurity he has already gnawed. In his first researches the scientist usually exhibits a rigid directness of aim. His gaze is fixed on the result and he does not focus the related information. His treatment is closely restrained to fact. As he develops his treatment becomes fuller. With confidence he can give sweep in his opinion. He begins to transform from a scientist into a natural philosopher (if he has the highest sort of mental quality) and reaches maturity after fifteen or twenty years of research. Between the ages of thirty-five and forty-five his work shows the finest balance of technical skill in handling fact and imaginative theoretical interpretation of fact. After forty-five years of age the factual and technical part of the research becomes a little less taut, and the imaginative theoretical interpretation a little less tightly aligned with fact. In old age these tendencies may develop further, until the consideration of fact becomes loose, and imaginative theorizing dissolves into speculation. One may not be surprised, then, if the chief creative period of Eddington and Jeans is finished. Eddington continues to publish many original papers, but the element of speculation in them increases. Jeans no longer publishes as many original papers as formerly. Their British successor is Professor E. A. Milne. He is not yet forty years of age and has not yet addressed himself to exposition. He has not written even one technical book: his work has been published in a long series of papers in learned journals. As he is still in his chief creative period his work shows the close attention to detailed fact, and this impairs the elegance of his arguments. The continual interruption due to the consideration of awkward relevant facts destroys so many expositions which would be so beautifully rounded, if only these awkward facts could be ignored!

Milne has made a very interesting contribution to the recent theories concerning the expansion of the universe. Astronomers have deduced from a study of nebular spectra that the distant nebulae are receding from the solar system. When an object is receding waves emitted by it appear to a stationary receiver to be of longer wave-length: the effect is frequently illustrated by the sound of the whistles of express trains. An observer on the bridge at a railway station will notice that the pitch of the whistle is higher when the train approaches than when it recedes after passing under the bridge. There is a sudden drop in tone, as if the engine, shrieking first in its rush on to the prey, had sunk into a contented hum, after successful capture. When light is the emitted wave the effect is exhibited by a deepening of colour, analogous to the lowering of tone in sound. A receding red light appears to a stationary observer to be redder than when it is stationary. As light travels extremely swiftly, 186,000 miles a second, a source of light must recede very quickly before the shift in colour towards the red becomes observable. It would have to recede several times as quickly as a rifle bullet before the effect would be noticeable. As all atoms of the same sort, such as hydrogen atoms, are known to emit light of a definite colour when stationary, a displacement of the colour towards the red or towards the violet indicates that the atom is receding or approaching the observer. A study of the light from the distant nebulae, that is emitted by the atoms of which they are constituted, shows they are receding from us, and the more distant they are, the more swiftly they are receding, because the shift in the colour is more pronounced.

In fact, the rate of the recession of the distant nebulae is proportional to their distance. Some of the near nebulae are approaching the solar system, but on the scale of the universe one has not to go far before one enters the outer space occupied by receding nebulae. This observation of the astronomers is remarkable. It suggests at first that the earth or the solar system is the centre of the universe, that the main content of the heavens is arranged round them, and that we are in a position of peculiar eminence in the celestial scheme. A conclusion so flattering to human vanity might have been extremely attractive if scientists had not from historical experience grown suspicious of anthropocentric views of the universe. They suspected the peculiarity of the earth's position according to this observation must be illusory. The first attractive explanation was suggested by Friedman and Lemaître. Some notion of their idea may be given by the following argument. The objects in the universe, such as stars and nebulae, have motions relative to each other. They may be regarded as moving in space. Movement is usually conceived as happening in a space which is given, and no one thinks about the behaviour of the space which contains the moving objects. Space itself is supposed to be fixed, given. Now suppose that space itself were in a process of change. The objects in the universe, stars and nebulae, would all have a movement compounded of two factors; for any object the first factor would be the ordinary movement noticeable in the relation to other stars and nebulae, and the second factor would be the movement superimposed on the object by the expansion or contraction of space itself. If space did not contract or expand this second factor would be zero, and the only movements of the stars would be those relative to each other. The second factor clearly would affect all objects, if it existed.

Observation shows that for distant nebulae the factor of motion common to all is immensely greater than the factor of motion relative to each other. The degree of the motion is proportional to distance only. All the distance nebulae at the same distance have equal observable motions, and the magnitude of their individual relative motions is negligible. Thus the factor of common motion is for distant nebulae vastly more important than the factor of relative motion. This indicates that their observed behaviour is much more influenced by one common cause, than by individual circumstances. One thing common to them all is the space which contains them. May not their common motion be due, therefore, to the common motion imposed on them by the space which contains them? The observed recession of the nebulae would be an observation, not of the individual movements of the nebulae, but of the expansion of space; the nebulae would be as lights on a moving ship, at rest relative to the ship but tracing out the path of the ship. They would be the markers of the expansion of space, fixedly attached to space itself, and serving as points for showing how space was behaving. Thus the observed recession of the nebulae informs of the behaviour of space, and not of the nebulae. If space were expanding this rate of recession of distant objects would be proportional to their distance. This can be illustrated by considering the behaviour of an expanding soap-bubble.

Suppose E N_1 N_2 are points on an expanding soap-bubble, such that the arc E N_1 equals the arc N_1 N_2. When the bubble has doubled in diameter the three points will be in the positions E_1 N^1_1 N^1_2. In the same period of time the lengths of the arcs between the points E and N_1 and between E and N_2 will have doubled. Thus E_1 N^1_1 will be twice E N_1, and E_1 N^1_2 will be four times E N_1, because E_1 N^1_2 is twice E N_2 and E N_2 is twice E N_1. Thus N_2 is retreating from E along the surface of the bubble twice as quickly as N_1. This argument may be generalized by saying that the rate of recession of a point N from a point E on the surface of an expanding sphere is proportional to the distance of N from E; if it is twice as far away, it recedes twice as quickly, and so on. Suppose now that E is the earth, N_1 is a distance nebula, and N_2 is another nebula twice as distant. Evidently the nebula twice

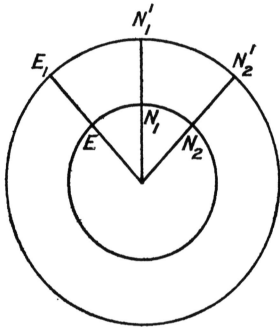

FIG. 18.

as distant will recede twice as fast, if the universe begins to expand.

If the universe is expanding the behaviour of the nebulae, which recede at a rate proportional to their distance, is explained. Is there any reason why the universe should expand? According to the theory of relativity the properties of space are not independent of its content. The stars and the nebulae are pouring out radiation in the form of light and other ether waves. It appears that if space does not expand, this pouring out of radiation would upset the stability of the celestial processes, but that if the universe is expanding the emittance of radiation is not a cause of instability. The effect is due to the pressure radiations, such as light, exert on the objects they strike. Any illuminated object is pressed by the illuminating light, as Maxwell deduced theoretically and Lebedev proved experimentally. Thus the continuity of the change of the universe, which is clear to observation, can be guaranteed, according to Lemaître, only by allowing that the universe is expanding. If there were no radiation, no expansion would occur, but a world full of radiation starts expanding as soon as the radiation can transform itself into matter. After matter is formed, there are complications due to gravitation, but the expansion depends entirely on the radiation pressure at places where the gravitational influences due to matter cancel out. Lemaître names these places neutral zones. The condensation of radiation into matter has in itself no direct effect on the stability of the universe, but condensations would necessarily induce rarefaction at the neutral zone and so a diminution of the pressure of the neutral zone, and this would induce expansion. The expansion could be started by the formation of the nebulae out of a diffuse primeval gas, or by the formation of stars out of the primeval gas with subsequent condensation into groups forming nebulae. Eddington finds the clue to the expansion of the universe in the nature of the electron itself. He has published calculations suggesting that the mass of an electron is a quantity related to the size of the universe and the number of electrons in the universe, in fact, that it is equal to the square root of the number of electrons in the universe divided by the radius of the universe. The number of electrons in the universe is about ten raised to the seventy-ninth power, i.e., 10^{79}, and the original radius of the universe comes out at 1,070,000,000 light years. Eddington has calculated from his formula the rate of recession of objects and obtains the value 528 (in kilometres per second per megaparsec). The observed rates are from 430 to 550. The very high rates of recession observed and calculated (the most distant nebulae are receding at 15,000 kilometres per second, which is one-twelfth the velocity of light) seem to require that the universe should be not more than 10,000 million years old, as the date of the beginning of the expansion may be calculated backwards. This is a thousand times less than the age required by the current explanations of the mode of the evolution of the stars. It is little more than the age of the earth according to geological evidence. Anti-anthropocentric caution would suggest that the earth is not a senior member in the population of the universe. The shortness of the age of the universe according to the space-expansion theory presents very serious difficulties. There are theoretical difficulties in the calculations due to assumptions of doubtful validity. Einstein and de Sitter have concluded that the theoretical calculations reveal no reason why space should expand rather than contract, the algebra does not show how the sign of the quantity representing the movement of the universe may be determined. Lemaître has calculated from the observed density of matter in the universe that any general process of condensation cannot be more than a few hundred thousand million years old. He asks, therefore, for a 'fireworks' theory of evolution. He considers the last 2,000 million years of evolution, which contain most of the life history of the earth, a period of slow evolution. If it is assumed that the universe always evolved at the present rate its age comes out too great. The present objects in the universe are the ashes and smoke of bright and very rapid fireworks. Lemaître believes the cosmic rays are the key to the problem. He explains that the energy possessed by all the cosmic rays in the universe is comparable with all the energy locked up in matter and the stars, perhaps it is equal to one-hundredth or more of the whole energy of matter. This is a very surprising conclusion. One would not have suspected that the rays which require such delicate technique for detection, which are very rare if very intense, could in any way bear a quantity of energy comparable with that locked up in matter. Consider the earth. Each atom is a packet of congealed energy, and an ounce of atomic energy would be enough to drive a liner across the Atlantic Ocean. If the whole mass of the earth could be converted into energy, each ounce of it would be enough to accomplish the equivalent of a liner crossing. Compare the stupendous stores of energy locked in the material of the earth with the energy of the starlight. The pale light from the stars is obliterated when a tallow candle is lit in nocturnal darkness. The cosmic rays reaching the earth are only one-tenth as intense as the starlight. How can the energy of all the cosmic rays in the universe be comparable with the energy of the billions of cosmic beacons, the stars; together with all the energy locked up as mass in matter? Because space is so empty. The volumes of empty space are so immense that if all the energy of matter were spread evenly through space the distribution would be tenuous. The cosmic rays, if they are really cosmic, are everywhere, are of the same intensity at nearly every point in space where there is no matter. So the sum of the energy of this tenuous radiation for the whole universe amounts to a vast quantity. Lemaître argues that the total original energy of the cosmic rays may have been larger than it is now, if the rays were produced before space began to expand, for it has been reduced in proportion to the expansion of the amount of space occupied by the rays. He considers the cosmic rays may have travelled around space for a period a hundred times greater than the age of the oldest known rays of light; those rays which left the most distant nebulae a hundred million years ago and to-day are registering themselves on the photographic plates of the most powerful astronomical telescopes. As the energy of the stars is the only sort comparable with that of the cosmic rays, the stars must presumably have been the parents of the cosmic rays. As the stars at present have atmospheres impenetrable to cosmic rays, they probably could not have produced cosmic rays while in their present condition. The rays must have been produced before the stars had atmospheres, and presumably the stars were born without atmospheres. Lemaître concludes the stars were born 10,000 million years ago without atmospheres, and that the cosmic rays are a chief feature in the formation of a star. He proceeds to some remarkable speculations. Cosmogony is atomic physics on a large scale, a large scale of space and time; why not also on a large scale of atomic weight? When an atom of radium disintegrates, penetrating rays named gamma-rays are emitted and resemble soft cosmic rays. May not the cosmic rays have been originally produced during the disintegration of a super-atom? a radioactive atom as large as the earth or a star? Uranium has the largest known atom and billions of uranium atoms would be needed to make a speck as large as a pin-head. Jeans has postulated the existence of radioactive atoms rather larger than uranium atoms to explain the source of stellar energy, but these would still be almost infinitely minute in comparison with an orange, and almost infinitely more minute in comparison with a star. Nevertheless, Lemaître's bold speculation is attractive, and his distinguished mathematical researches entitle him to unusual freedom in speculation. He regards a star as the offspring of one of these radioactive super-atoms. When the super-atom disintegrated it emitted a vast shower of particles and waves. The wave radiations were a constituent of cosmic rays, which are, in Lemaître's picturesque expression, glimpses of the primeval fireworks of the formation of a star from an atom coming to us after long journeying through space. The energy of cosmic rays is high, but perhaps not high enough to be the product of such a super-atomic cataclysm. Their original energy would have been higher, however, as their frequency (and hence energy) would increase in proportion to the expansion of space. The existing cosmic rays may for this reason be only of one-twentieth their original energy. Recent researches on the cosmic rays have increased the interest of Lemaître's theory of their origin. If they were emitted from a disintegrating super-atom they were presumably accompanied by showers of electrons and helium nuclei, as the gamma-rays from radium are so accompanied. Perhaps they were accompanied by new rays of greater masses and charges. He calculated that the momenta of the particles in particulate rays would be reduced by expansion of space in about the same ratio as that of the light-units in the cosmic rays. Cosmic ray research is increasing the possibility that they may contain particles besides wave-radiations, or may be constituted entirely of particulate radiations. The recent discoveries of the neutron and the positron are a reminder of the possible existence of still more sorts of radiations. There is evidence that at least 50 per cent of the cosmic rays are positrons.

In Lemaître's theory, the universe need not be more than ten times as old as the earth. In the beginning it was concentrated in one super-atom. The whole universe was no larger than the earth or perhaps the sun. As this super-atom disintegrated it flung out particles and radiations which caused space to expand, and after 10,000 million years space has expanded to the present size. Some of the fragments of the exploded primeval atom will be heavy enough to prevent the escape from them of adjacent particles and débris; by gravitation they will form ordinary stars, and clusters of stars. The cloud of loose particles, the electrons and simple atoms known to us would be spread as a gas through space, and the nebulae would form out of this gas by the gravitational condensation processes investigated by Jeans. The theory requires an extension of the knowledge of atomic structure. Methods of closely packing electrons to form new sorts of atoms of enormously high atomic weight would have to be discovered. The ordinary atoms, such as those of iron, are very spacious. New sorts with few interstices must be imagined and their existence shown to be possible.

The conception of the universe as the unfolding of a compact object no larger than the earth has a certain biological quality. The growth of the seed or the egg into the vastly larger adult has an analogy to the Lemaître conception of an unfolding world.

That is a summary of the ideas of one of the founders of the theory of the expanding universe. Milne has suggested an alternative explanation of the uniform and proportional recession of the distant nebulae. He starts from purely classical ideas. He does not even assume that space is curved, and can quite dispense with the thrills of an expanding space. He needs little more than the old, fixed, Euclidian space of the school books and ordinary human affairs. Suppose that in the beginning the whole of the infinite fixed familiar Euclidian space were empty, except in one quite small spherical region. Suppose all the stars and nebulae were within this sphere, and moving about at random. The velocities of relative movement would vary from zero to the maximum possible: the velocity of light. Some would be moving towards the centre of the sphere and others outwards. In the lapse of time those outwardly-moving stars near the surface would pass through the surface of the sphere and journey into space. The inwardly-moving stars would approach more nearly the centre of the sphere. After a further interval these would have reached the centre and passed on towards the other side of the sphere, they would have become outwardly-moving objects. After a still larger interval the majority of the stars would be journeying on through the empty space, and those stars which started with the highest velocities would have gone farthest. The most distant objects would be receding most swiftly and receding at a pace proportional to their distance, because their distance is proportional to their pace. This is the condition observed in the distant nebulae. Thus the remarkable fact of the swiftly-receding nebulae may be explained as a simple example of diffusion. It is analogous to the diffusion of a bubble of gas released in a vacuum. The most swiftly-moving of the molecules in the bubble will penetrate farthest into the vacuum, so that the molecules most distant from the original bubble will be receding fastest. The explanation will hold even if space is not originally empty. An initial unevenness of distribution of objects is sufficient to produce the phenomenon, if sufficient time and space are allowed. In the more refined statement of his idea Milne introduces new applications of the theory of relativity. He avoids a difficulty concerned with the concept of 'cosmic time', essential to the space expansion theory; and extends the principle of relativity, which in its original form applied to the laws of nature, to the objects in the universe itself. He deduces that the universe must appear the same to all observers, whatever their position, if local irregularities are disregarded. Every observer can regard himself as the centre of the universe by choosing his time-axis suitably, away from the time-space origin. "The world is then perfectly egocentric at all points, and the moving picture of the world as made by any one observer is identical with that made by any other observer."

Milne's theory of the constitution of the stars is even more interesting. By a remarkable mastery of the mathematics and the logic of a large combination of physical and astronomical facts he has shown the various types of stars fit into a comprehensive scheme. In particular he has suggested the very important rôle of novae, the stars which flare up brilliantly and rapidly die down again, in the process of stellar evolution. The theory of degenerate matter, first cultivated by R. H. Fowler, is an important constituent in Milne's account of novae. The peculiar properties of degenerate matter may be explained according to the quantum theory. The ultimate particles such as electrons may overlap each other but cannot be both in the same place and in the same state at the same moment. This exclusiveness of state implies that there is a limit to the closeness with which electrons may be packed. A portion of space may be so filled with electrons in a variety of states that no more may be packed in, because no more states are available to serve as entrance tickets. Ordinary atoms are, in fact, examples of portions of space packed with the maximum number of electrons according to the principle of the exclusion of similar states. Imagine a super-atom constructed not of a vast package of ordinary atoms, but of electrons and protons arranged as a huge unit, the structure being dependent mainly on the operation of the exclusion principle. The super-atom might be as large as a star. E. C. Stoner has calculated what its density might be, if its volume were the minimum compatible with the observation of the exclusion principle, i.e., with the space-requirements of electrons. He finds that the density is proportional to the square root of the mass of the star, and is 3,850,000 times that of water for a star of mass equal to that of the sun. Portions of space containing the maximum number of electrons and protons would be more approximately compared with super-atoms than with portions of ordinary matter such as the earth, as Fowler explained. (This notion of the degenerate star conceived as one atomic unit is reminiscent of Lemaître's original atom-universe.) In degenerate matter, which is extremely dense, and in which electrons are as close as the mutual exclusiveness of their states will allow, the electrons have no room in which to do anything except exchange places and states. There is no room to interact with external agents. This has a curious consequence. Though extremely dense, degenerate matter is extremely transparent. A ray of light or radiation goes through it with ease - because all of the electrons, closely and densely packed as they are, are unable to interact with the penetrating ray. If they did react, their state would change, and by hypothesis all available states for electrons are already occupied within that region. Light and heat within degenerate matter rapidly leaks away. Consider what this may mean in an ordinary star. The star is mainly gaseous, as its density indicates, for this is usually about the same or less than the density of water. The gaseous structure is partly supported by the pressure of the intense radiation inside. The outer layers of gas are supported by the beams of intense radiation from the hot centre. As time passes the pressure of the radiation becomes feebler. A moment comes when it is no longer sufficient to preserve the ordinary gaseous star structure. The radiation pressure has decreased more rapidly than the gravitational pressure, and the latter has assisted in forming a little portion of degenerate matter in the centre of the star. What happens after this degenerate core has been produced? The transparency of the degenerate material allows a sudden outrush of radiation. This presses fiercely on the outer gaseous layers which are more resistant to it. The outer envelope of the star is blown off, and appears as an expanding sphere of flame. The sudden extension of shining material, and the energy released by gravitational condensation when the inner material collapses into the core of degenerate matter, increases the star's luminosity. Previously it might have appeared quite faint, now it blazes into sight as a nova. Meanwhile, the loss of radiation has still further weakened the internal structure of the star, so more matter becomes degenerate, and the inner part of the star condenses while the outer parts are flying outwards on their impelled journey. This is Milne's theoretical explanation of the production of a nova. He regards the nova outburst as a climacteric occurring in the life of every star, or nearly every star. It is a general characteristic of stars. This is consonant with the remarkable frequency of novae. In our own local universe of the Milky Way about one new nova is seen every year. In other local universes such as the Andromeda Nebulae, novae appear more frequently. As the universe is at least tens of thousands of millions of years old, the number of novae in universal history must be of the same order as the number of stars. Does this not suggest that all stars pass through the nova stage at some date in their lives? The detailed observational knowledge of novae offers remarkable agreements with Milne's theory. Within ten years a nova returns to approximately its original luminosity, but spectral evidence shows that its surface is much hotter than before. If the same light is coming from a hotter surface the area of the surface must be less than before. This implies that the star has become denser. Actual observation of novae has indicated that the star may be one thousand times as dense after the outburst as before. The rapidity and brightness of a nova outburst is a product and a measure of the degree and rapidity of the cataclysmic condensation of the interior of the star.

Novae are of great natural philosophical importance because their changes in time are directly observable. One man may see a nova pass through many conditions and increase its density one thousandfold during a decade of his scientific life. He sees a star definitely evolving. The evidence of evolution from the other stars is strong but not so direct. The hundreds of sorts of stars may be arranged in a continuous sequence according to properties, which strongly indicates they are changing from one sort into another, but these changes are too slow to be observed directly. The fact of the change is inferred from the sequence but is not itself visible. In novae the changes are swift and directly observable, so they provide the evidence for stellar change as a fact, and not merely as an inference, however strong.

The observation of the production of stars of material hundreds of times denser than water, and therefore many times denser than any material known on the earth, increases the significance of the existence of other sorts of dense stars. The story of the

Companion of Sirius is famous, but it continues to grow as an Arabian or a medieval tale grew when the appropriate cultures flourished. The tale of science is the Homeric story of modern culture. It grows because the civilization which bore it is growing, and unlike the Homeric story is not finished because the existing process of a scientific civilization is not yet finished. The calendar and civilization of Ancient Egypt were regulated by the rising of Sirius. The present name of the star is Greek. In 1718 Halley published his great discovery that Sirius and others of the principal fixed stars were not quite fixed. He had found they had moved slightly but systematically through the centuries by comparing their positions as recorded by Timocharis and Aristyllus nearly 300 years before Christ, and by Hipparchus 170 years before them, with the positions they occupied in his own day. He wrote that

> "these Stars being the most conspicuous in the Heaven are in all probability the nearest to the Earth and if they have any particular Motion of their own it is most likely to be perceived in them, which in so long a time as 1800 years may show itself by the alteration of their places, though it be utterly imperceptible in the space of a single Century of Years."

This was a major discovery in the history of science. Hitherto the fixed stars had been the symbol of finality. They were the eternal setting of the affairs of the earth. Human life was a puny flickering before myriads of eternal constancies. The conception of fixity in the external world diminished the strength of belief in the power of change, and reinforced social reaction besides exalting constancy. Halley's destruction of the belief in the fixity of the stars was an important liberation of the human spirit. An ancestral conservatism was destroyed, and henceforth confidence in the possibility of change was increased, if even the apparently eternal stars change and evolve. Why was the discovery of the movement of the stars reserved to Halley? He must have been sensitive to the idea of change if he so acutely noticed it in the slight discrepancies in the records of the fixed stars. The great minds of his epoch, Harvey with his circulation or movement of the blood, Galileo with his laws of motion, Newton with his calculus of motion, were concerned with the philosophy of change. Why should the phenomenon of change be at the centre of their attention? Because their social life was changing. The slowly-accumulated mastery of the production of the means of life had by the sixteenth century provided a capital for an extension of human adventure. The possibility of change from poverty to riches had rather suddenly increased, and a class of able opportunists arose which employed the new more convenient forms of commerce, industry and transport for self-aggrandisement. The turmoil of the economic and social movement named the Reformation made the best minds of the day sensitive to the idea of motion.

Halley's discovery of the motion of the fixed stars was the foundation of evolutionary theories of the stellar universe. It was the greatest observational contribution to the theory of cosmogony, now a part of the culture of every educated man. The contribution of Sirius to the instruction of humanity did not end with the announcement to the ancient Egyptians of the imminence of the rising of the Nile and guarantee of harvests and life, nor with the inspiration of Halley to the discovery of dynamic cosmogony. Sirius is the most brilliant and therefore probably one of the nearest stars, so it could not be expected to have more instructive characteristics, after having provided so many. To be both near and queer would have been extravagant, yet Sirius has proved extravagant. In 1834 Bessel began to suspect that Sirius not only moved but wobbled. He found a periodicity of fifty years in the star's movement, and suggested that the star might be double, one component being bright and the other dark, and the pair rotating around a common centre of gravity itself in uniform motion according to the laws of mechanics. This would fit the observed facts. Bessel believed the phenomenon would reveal important knowledge concerning the structure of the universe. He thought novae might be explained by the mutual revolution of a bright and a faint star, and mentioned that Tycho's nova seemed to have shown peculiarities explicable by the passing over and losing of a bright body. Milne's theory of novae has given remarkable significance to Bessel's intuition that the explanation of the irregularity of Sirius would give new knowledge concerning the physical characteristics of the universe, and might be connected with novae.

In September 1861 Safford assigned a position to the hypothetical Companion of Sirius. In January 1862 Alvan Clark was testing the object-glass of a large new telescope made by his firm. He happened to direct the telescope towards Sirius in order to see the quality of the definition. Near the star he noticed a hitherto-unrecorded point of light. It was close to Safford's position for the hypothetical Companion, and indeed it was the Companion. The cause of the irregularity of Sirius was not dark, but visible though faint. A study of its motion showed that its mass was almost equal to that of the sun, whereas the mass of Sirius is nearly two and a half times that of the sun. The faintness of the Companion was ascribed to low temperature, and presumed redness.

In 1914 W. S. Adams discovered that the Companion was white and not red; it was evidently bright. Its faintness could be due only to small size, and the size was calculable from the distance of the star and the spectroscopically measurable brightness of its surface. Its diameter proved to be roughly one-fortieth that of the sun; which is a star of about equal mass. The density of the Companion calculated from this dimension and mass is 60,000 times that of water, about one ton per cubic inch. In 1914 this result seemed absurd. The heaviest metals known are only about twenty times denser than water. Ten years later Eddington explained that very dense materials were not inconceivable, as atoms are roomy objects and might be reduced in size by suitable conditions. According to the Rutherford model of an atom, a very small nucleus is surrounded by distant crinolines or rings of revolving electrons, leaving immensely more space than material in the atomic structure. If the outer rings of electrons could be partially or entirely removed, the size of the atom would be enormously reduced. Under the high temperatures and pressures in the centres of the stars the atoms would collide violently enough to knock off each other's outer rings of electrons. The reduction in atomic size would allow the constitution of super-dense material. This explanation made the existence of super-dense stars theoretically probable, and the absurdity of the density of the Companion of Sirius disappeared.

Eddington and others pointed out that this happy explanation might be confirmed by an entirely independent method of observation resting on the theory of relativity. Rays of light passing by massive bodies are slightly distorted. The degree of distortion is proportional to the mass divided by the radius of the star, so dense stars should show the effect more markedly than less dense stars. For instance, a passing ray of light should be distorted by the Companion of Sirius thirty times as much as by the sun. As the demonstration of the existence of the sun's Einstein effect on passing light from a star was one of the chief experimental proofs of the theory of relativity, the possibility of measuring the high density of the Companion of Sirius through this Einstein effect was particularly interesting. Eddington calculated that the Companion should appear to reduce the speed of the light passing it by 20 kilometres per second; in technical language there should be a displacement of the lines in the spectrum towards the red, equivalent to a reduction of the speed of light by 20 kilometres per second. The observations were difficult because of the proximity of Sirius, which is 10,000 times brighter than the Companion. In 1925 Adams succeeded in overcoming the difficulties, and found the shift to be 19 kilometres per second, in marvellous agreement with Eddington's theoretical prediction. In the earlier theoretical studies of the possible states of the interior of small white stars such as the Companion of Sirius the stripping or degenerating of the atoms that allowed the existence of very dense matter was attributed to the destructive effect of extremely high temperature. The white dwarf stars were supposed to be extremely hot, with centres at temperatures of hundreds or thousands of millions of degrees. The later studies showed that the ordinary conceptions of temperature lost their meaning because a portion of matter in the degenerate state must be conceived as one unit, and not a collection of discrete particles. The conception of temperature is statistical, it is derived from the average behaviour of a large number of particles, and does not apply to one particle. Thus the white dwarf stars need not have extremely high central temperatures. Milne calculates that 15,000,000 degrees is a reasonable central temperature for a white dwarf. This is much lower than the central temperature of ordinary stars such as the sun. The transparency of degenerate matter allows heat to escape and spread evenly throughout the body of the star. The high temperature of the surface of a white dwarf is due to the proximity of the hottest material. In an ordinary star the temperature rises gradually from the surface to the centre, so the surface may be cool while the centre is extremely hot, the surface being far from the hot centre. In the white dwarf the comparatively cool degenerate material comes almost to the surface. The thin layer of gas on the surface of a white dwarf acts as a blanket or cosy. Milne has compared it with the asbestos coating used for conserving the heat of the water in a hot-water tank. The gaseous envelope is less transparent to radiation, so it keeps the heat in. The difference in properties between the gaseous material and the degenerate material is sufficient to define the condition of degenerate matter as a new physical state. Water exists in the states of ice, water and steam. Milne regards degeneracy as a fourth physical state analogous to these and also dependent on the temperature and pressure. He has worked out a theory of the structure of stars as possible configurations of matter in two states, the core of the star being in the degenerate state and the upper layers in the gaseous state. One could imagine an analogous problem of a body constructed of a ball of water surrounded by an atmosphere of steam. The balance between the size of the liquid core and the gaseous envelope would depend on conditions of temperature and pressure. Milne finds that his gas-

and-degenerate-matter stars might exist in several ranges of size and properties, and in part of the range there is a duplication of possibilities. Under certain conditions the factors determining whether a star shall be in one range or the other become indefinite. The star might pulsate from one range to the other. This is Milne's explanation of the existence of the Cepheid pulsating stars, those very luminous stars which pulsate in luminosity as regularly as clocks. It is perhaps the only promising explanation of their mechanism which has been suggested. The gas-and-degenerate-matter theory gives plausible accounts of novae, white dwarf stars and Cepheid pulsating stars. Planetary nebulae may be included, for they are probably the post-eruption condition of large novae which expelled sufficient gas to form the spherical attendant nebulae. Milne finds the explanation of the cosmic cloud, the thin scattering of atoms throughout the whole of space, in the expulsions of gas from novae. His nova theory of double stars is interesting, especially in relation to Sirius. He finds that a star passing through its phase of nova eruption may be particularly sensitive to rotational splitting. The collapse of much of the material and the mass of the star into the core will upset the moment of inertia. The central region will have to rotate at an increased rate in order to conserve the angular momentum. This may cause the star to be torn into two parts, and the parts may not necessarily be alike. One may be mainly of degenerate matter and the other of gaseous matter. This would explain the origin of the Sirius pair. They were formed by rotational instability attendant on the nova eruption of the parent star. Milne remarks that if this is correct Bessel's intuition of the connection between the Sirius double star and novae was correct and his belief that the Sirius phenomenon would be specially instructive concerning the structure of the physical universe was justified. Milne sees the universe as a concourse of stars condensing into denser stars in virtue of instability and waning light-pressure, and at the same time generating a nebula by the expulsion of their surface layers of gas. He contrasts this with Jeans' view that the universe started as a diffused nebula which condensed into stars through gravitational condensation, the particles of gas gradually attracting themselves together into denser and denser gaseous units.

CHAPTER V

COSMIC RAYS AND THE POSITRON

I

THE problem of the nature and properties of the cosmic rays is the most interesting in the newer branches of physical research. These rays have various names. German workers often refer to them as ultra-rays, American workers as cosmic rays, and English workers as penetrating rays. The general reader will find the name 'cosmic rays' the most useful because it suggests their wide scope and relations, though as a scientific name it is not severely correct because the rays are not quite certainly cosmic. The latest researches have proved that the rays almost certainly are of cosmic origin, but the scientist prefers to use names which contain no implication of properties not completely proved. The name 'cosmic' was used for several years before the certainty of cosmic origin became high, and some of the severely critical scientists did not like the use of a term which suggested a description of more than the observed facts. The name 'cosmic rays' will probably become established because, like another scientific name chiefly familiarized by America, the word 'vitamin', its romantic quality has a general appeal. The name 'vitamin' was proposed when those substances were believed to be derivatives of the chemical amines. Some of them are now known not to be amines, so the word is not strictly descriptive of their properties, but it has superseded

122

the severely scientific name 'accessory food factor' which does not include more than the observed properties in its description of these substances. Names such as 'very penetrating radiation' and 'accessory food factor' exhibit the discrimination of the pure scientist who avoids names however attractive because they have other associations and therefore allow old emotions to distort present intellectual judgment. When a phenomenon is thoroughly understood and old a slightly inaccurate but convenient name is harmless. 'Vitamin' is now useful and harmless.

If a starting-point in the study of cosmic rays must be given, the most suitable is perhaps made by the researches of Rutherford and Cooke in 1902. Their address to the American Physical Society on December 31st of that year contains the most definite of the early contributions to cosmic ray research. The researches of Geitel in 1900, and of C. T. R. Wilson in the same year, contain pregnant observations, as Geitel had found that the air in closed vessels in places apparently free from radioactive substances had a definite electrical conductivity, and Wilson remarked that the electrical conductivity he had observed in air in closed vessels might be due to the action of an external agent, and even suggested that it might be a radiation coming from outside the earth. But the first clear assertion of the existence in the air of a very penetrating radiation is due to Rutherford and Cooke. It was made with the direct simplicity and confidence characteristic of Rutherford. He had recently been investigating the γ-rays emitted by radioactive substances. These are a wave-radiation of great penetrating power. As with other radiations, their presence is detected by their ability to increase the electrical conductivity of air in a closed vessel, due to the liberation of electrons from the atoms of the air gases. The properties of the γ-rays from radioactive substances could be studied by this method only when the behaviour of closed chambers of gas free from γ-rays was already known. Hence the conductivity of the air in such a closed chamber when free from the γ-rays of radioactive substances must be known. Rutherford and Cooke discovered that great thicknesses of lead noticeably reduced the conductivity. They cast squares of lead 1 inch thick which could be built as a leaden box. A 2-inch layer of lead around the chamber reduced the conductivity by 30 per cent. When the layer was removed the rate returned to the original value. They wrote in the *Physical Review* for 1903:

"These results show that about 30 per cent of the ionization inside a closed vessel is due to an external radiation of great penetrating power. This radiation appears to come equally from all directions and is probably due to excited activity on the surface of the room in which the observations were made. These effects could not be due to the presence of thorium or radium in the laboratory, for similar results were observed in the library which was free from all possible contamination by radioactive substances."

On the next page of the *Physical Review* there is an account by McLennan and Burton of their experiments on the conductivity of gases in closed vessels. They surrounded the vessels with layers of water. They concluded that

"the effects up to the present may be explained by supposing the ionization to be caused by a radioactive active emanation sent off from the metals, it has been found that part of the conductivity cannot be accounted for in this way, but must be attributed to an active agent external to the receiver. . . . It is evident that the ordinary air of a room is traversed by an exceedingly penetrating radiation such as that which Rutherford has shown to be emitted by thorium, radium and the excited radioactivity produced by thorium and radium."

Rutherford and Cooke, and McLennan and Burton, supposed that these very penetrating rays present everywhere on the earth's surface were due to radioactive matter distributed through the material of the earth and the atmosphere. McLennan experimented with vessels in Lake Ontario to escape the rays coming from the material of the earth and in 1907 wrote that the number of electrons liberated in one second in an aluminium vessel could be reduced to 4·8 per c.c.

These early experimenters were concerned with the problem of radioactive contamination. The wide distribution of radioactivity made the production of apparatus free from radioactive substances difficult. As they had experience of the effects of traces of radioactive matter in the metals of their instruments and had learned of the necessity to use apparatus made of material not contaminated with radioactive matter, they naturally attributed the very penetrating rays in the air to radioactive material distributed in it. They considered the possibility that all matter is in some degree radioactive. Schuster had remarked in 1903 that all materials hitherto known shared the same physical qualities. The difference between elements or substances is in the degree of these qualities. For instance, all substances are magnetic, though in varying degrees. Solids are slightly liquid, and liquids are slightly solid, and so forth, so all substances may be expected to be radioactive. Most of the early workers discarded this view, and as a working principle it was not at that time of use. Modern wave-mechanical theories of matter show that many sorts of matter are radioactive, but in a degree are often too small to be detected by present experimental technique. The slight ionization of gas in closed chambers protected by screens of lead or water was attributed not to the spontaneous radioactivity of the material of the vessels and the enclosed air but to contaminations mixed in them.

The existence of very penetrating radiations in the air was demonstrated by measuring the electrical conductivity of air caused by them. It followed the investigation of the properties of radioactive radiations. The idea of penetrating radiations was in the mind of Rutherford and other investigators because certain radioactive substances give such radiations. They looked for evidences of other radiations resembling the radioactive radiations in which they had been interested. Their view of the explanation of the nature of the very penetrating radiations was a laboratory view: it saw the origin of the radiations as proximate. In laboratory experiments the source of an agent is usually within the room. It may be a tube or a flame or a magnet. The laboratory experimenter in his explanation of phenomena tends to assume the source of an activity to be within his room or near to the effect; and this tendency is observable in the early theories of the nature of the very penetrating radiations. These observers were not primarily interested in the nature of the electrical conductivity of the air. They were interested in radiations, and consequently interpreted the ionization-chamber phenomena as due to radiations. The history of physics continually exhibits how an idea may be fruitful in a particular period when it is applicable, and how its associations may limit the extent of its applicability. In 1902 very penetrating rays were associated with proximate sources, so the very penetrating rays in the air would be assumed to come from proximate sources.

Other investigators brought a different mental

environment to the research. They were not interested so much in radiations as in the properties of air and in meteorology. This group of workers was interested primarily in the electrical and general properties of air and gases. Geitel and Wilson were in this group. They were interested in the electrical behaviour of gases first, and in radiations only as a means to explain the electrical properties of gases. Thus Wilson, for example, who had already shown that the rates of loss of charge of enclosed gases was approximately proportional to pressure and density, casually suggested the existence of a penetrating agent to explain ionization phenomena in closed chambers of gas, but he did not fix the attention on the agent because he was more interested in the effects. As the students of atmospheric ionization, the electrical conductivity of the air, were not primarily interested in radiations, they did not see in their observations particularly striking evidence of the existence of radiations and therefore were not the first to demonstrate the existence of very penetrating radiations. Their researches had a different ideological orientation which proved to be fruitful in another aspect of the cosmic ray problem.

The students of the properties of the atmosphere, meteorologists, bring to the interpretation of phenomena a set of general ideas different from those of the laboratory worker. They must often consider the properties of the complete atmosphere in the search for the interpretation of atmospheric phenomena. The atmosphere exhibits electrical phenomena such as thunderstorms, in which charges of electricity are exchanged between the clouds and the earth. There is a steady difference in electrical voltage between the earth and the air which must be preserved by some general agency, and the meteorologist naturally tends to think of agencies capable of acting simultaneously throughout the world's atmosphere. This causes a tendency to interpret phenomena by agencies from distant rather than from proximate sources.

In 1905 Campbell and Wood remarked that there appears to be a definite relation between the periodic variation in rate of ionization of air and other gases in closed vessels and the periodic variation in the intensity of the electric field near the earth's surface. They found the ionization maxima occurred at 8 a.m. and 10 a.m., and 10 p.m. and 1 a.m., while minima occurred at 2 p.m. and 4 a.m. The maxima of the voltage in the air are roughly parallel, while the minima are remarkably constant and occur fairly exactly at the same hours. In 1906 O. W. Richardson suggested that this connection between the periodicity in the degree of ionization in closed vessels and the periodicity in the variation in the intensity of the electric field near the earth's surface might readily be explained by the theory of the conduction of electricity through gases, if the ionization is assumed to be caused by radiation coming from outside the earth. As the variations in the electric field near the earth's surface were known to be connected with the sun, the variations in ionization in closed vessels might also be connected with the sun, so Richardson suggested that the ionizing radiations came from the sun. He remarked that the rays would to some extent be absorbed by the earth's atmosphere and 'they will therefore be more intense further away from the earth's surface'. The effects that had led Richardson to this view were shown later to be due to intermediary meteorological phenomena rather than to cosmic rays. In 1909 Gockel made balloon ascents to a height of 4·5 kilometres, taking with him ionization chambers. He discovered that the intensity of ionization did not decrease at the rate to be expected if it was due to radioactive substances on the earth. At 4·5 kilometres the intensity was greater than at the surface. This was the most important discovery since the researches of Rutherford and of McLennan.

The first adequately sensitive ionization chamber for observing cosmic rays was constructed by Wulf in 1909. It consisted of a vessel containing an electroscope of special design. Two parallel filaments in contact were given an electrical charge so that they repelled each other and bowed outwards. The distance between them could be measured by a microscope and gave the size of the charge. As the electroscope lost its charge through the ionizing of the surrounding air the two filaments fell closer and closer together. By making periodic observations the rate of discharge of the electrometer could be determined. Wulf showed that if the very penetrating rays were similar to the penetrating rays from radioactive substances the latter could not be absorbed by the air according to the ordinary laws of ray absorption. Hess showed that the absorption of all known radioactive radiations obeyed the ordinary laws. Thus the properties of the very penetrating radiation appeared to be irreconcilable with a radioactive origin.

Between 1911 and 1914 Hess, Wulf and Kolhörster made a number of balloon ascents with improved apparatus, Kolhörster ascending to a height of 9 kilometres; at that height he found the rays to be seven times as intense as at the earth's surface. They conclusively showed that the intensity of ionization decreases up to a height of about 1 kilometre and then increases, becoming equal to the earth-surface value between 1 and 2 kilometres. Hess considered his experiments showed the ionizing effects must be due to rays coming from outside the earth, as suggested by Richardson. At that time Gockel and Wulf were still inclined to believe the very penetrating rays were due to radioactivity.

The return of Halley's comet in 1910 provided the opportunity of observing whether any penetrating ray effects were caused by radioactive materials in the comet's tail. The earth swept through the tail, and conceivably its atmosphere might receive an injection of radioactive particles. Some observers believed they had evidence of the effect, but their observations were probably due to accidental coincidence, or to meteorological or instrumental disturbances.

Other early suggestions concerning the nature of the very penetrating radiations were those of Birkeland and Kolhörster. Birkeland suggested the rays consisted of swift electrons flung out of the sun's spots or faculi, while Kolhörster suggested the rays consisted of electrons emitted from the sun through thermal effects, the same effects which cause the hot filament of a thermionic valve to emit electrons.

Electrons emitted from the soil may cause noticeable ionizing effects at a height of many metres.

The next notable advance occurred when physical research had begun to recover from the distractions of war.

In 1914 Millikan had become interested in the rays but was not able to work much on them until after the war, in 1921 and 1922. He aimed at the measurement of the rate of ionization on mountains, in deep lakes, and very high in the atmosphere by free balloons carrying very light automatic electroscopes. He designed an instrument that could make a continuous record during flight of the rate of electrical discharge, temperature and barometric pressure. This triumph of the mechanic's art weighed 7 oz. It was attached to two balloons each 18 inches in diameter when deflated. In order to make an ascent the balloons were filled with gas and released. They continued to rise until one of them burst, and then the other acted as a parachute so that the rate of descent was retarded and the apparatus landed gently without damage. The inflated balloon was conspicuous as it lay on the ground, and attracted attention, so that the record was not lost. In the most successful flight the balloons with the instrument rose to a height of 15·5 kilometres and landed at a distance of over 100 kilometres. The temperature at the maximum height was − 60° C. and the time of ascent 115 minutes, and of descent 76 minutes. The total time of flight was therefore 3 hours and 11 minutes. The self-recording electroscope showed that the intensity of the rays steadily increased up to the maximum height, but the rate of increase was much less than would have been deduced from Kolhörster's observations up to 9 kilometres. This showed that the penetrating power of the rays must be much more than previous observations suggested, as the large extra layer of air between 9 and 15·5 kilometres had not reduced the intensity by the expected amount. Nine-tenths of the atmosphere by weight is below 15·5 kilometres, so the ascent left a layer of only one-tenth unexplored.

The next advance was made by Kolhörster in 1923. He attempted to measure exactly the rate at which the rays were absorbed in materials other than air. The absorption coefficient, which is the measure of penetrability of the rays, was deduced in previous experiments on the assumption that the rays came vertically down through the atmosphere. This route was rather arbitrary in a layer of material many kilometres thick. With thinner layers of denser material the route of the rays was less important.

The absorption of rays is proportional to the quantity of matter through which they pass. A barometer tube shows that a column of air as high as the top of the atmosphere is equivalent to 76 centimetres, or rather more than 2 feet, of mercury, which is equivalent to about 30 feet of water. Thus a layer of pure water 30 feet deep would double the shielding of an electroscope already protected by the whole depth of the air of the atmosphere. Kolhörster measured the absorbing effect of ice in Alpine glaciers under which ionizing chambers were placed, and he made the first measurements of absorption by water, by sinking a chamber 12 metres in the lakes near Berlin. He found that 1 metre of water reduced the intensity of the rays by 25 per cent, which gave a penetrating power less than half that he had deduced from his early balloon ascents and in better agreement with the records obtained by Millikan's free balloons. He remarked that "one inclines more and more of late to the view that the penetrating rays are a phenomenon the origin of which is to be found in the cosmos". Millikan and Otis made measurements of the absorbing power of thick lead screens on the top of Pike's Peak, and concluded that the rays were either not cosmic or of penetrating power greater than that found by Kolhörster.

The measurements of the absorption by water became more and more refined. Millikan and his colleagues sank electrometers in the water of snow-fed lakes in North and South America. The water of these lakes was exceptionally pure and free from radioactive contamination. Glacier ice and rocks might always contain radioactive material and disturb electrometers near them. In 1925 Millikan found that an electroscope sunk below 18 metres or 60 feet in Muir lake showed no further decrease in its charge. Muir lake is 3,590 metres or 11,800 feet above sea-level. Similar experiments in another lake 300 miles away but at a height of only 2,060 metres, 6,700 feet, showed similar results, except that the rays appeared to penetrate only 54 feet. Six feet of water is exactly equal in absorbing power to the layer of air in the atmosphere between 6,700 feet and 11,800 feet! While Millikan was investigating the rays in the mountain lakes of North America and Kolhörster in Germany, Myssowsky and Tuwim, in Lake Onega in North Russia, were measuring the absorption of the rays by water and obtaining parallel results. Millikan continued his observations in the lakes of the Andes in South America, on Titicaca, and on Miguilla in Bolivia, with concordant results. He also took sea-level measurements regularly during the sea voyage from Peru to Los Angeles, and found no change in intensity with latitude. The observations in North and South America and in Europe seemed to show that the rays come into the earth with equal intensity in all directions. Millikan could find no reliable relation between intensity and place on the earth, with the position of the sun or stars. He inclined to the view that the rays came from the depths of space, perhaps from the diffused matter. This would explain their apparent uniformity. The universe appeared to be uniformly lighted with the cosmic rays. There were no beams and shadows.

Besides showing that the rays could penetrate many metres of water Millikan found evidence that they were not homogeneous, that the rays consisted of a mixture of rays of different penetrating powers, so that they exhibited a spectrum. From statistical analysis of his observations he concluded that some of the rays had energies equivalent to those necessary for the construction of atoms of helium, iron and other elements out of simpler atoms. His discussion of the rôle of cosmic rays in atom-building received much notice.

At this time the very penetrating rays were generally assumed to consist of photons. They were supposed to be ultra short wave-radiations. The supposition was due to previous experience. Penetrating rays hitherto had always been wave-radiations such as X-rays, or the gamma radiations from radioactive substances. The swiftest known electrons were absorbed by a few millimetres of metal foil, so the penetration of 60 feet of water, equivalent to 2 yards of lead in stopping power, could not with probability be ascribed to electrons.

Exact measurements of absorption continued to be made by many workers. Steinke in 1928 sank an apparatus weighing half a ton to a depth of 40 metres in the Masurischen Lakes in East Prussia. He detected rays at that depth, and his measurements agreed with the measurements of Millikan and Cameron made in the same year. The deep-water investigations were greatly extended by Regener. In 1931 he announced that his observations on Lake Constance showed the presence of rays which penetrated to a depth of 236·5 metres, or 770 feet. The lake is in one part 820 feet deep. His apparatus consisted of an electrometer filament which was caused by an electric lamp to throw a shadow once an hour on a photographic plate. The electrometer was connected to a steel ionization chamber of 33·5 litres capacity with walls 1 centimetre thick, filled with carbon dioxide at a pressure of about 420 lb. per square inch. The high pressure and density of the gas increased the stopping power and hence the quantity of ionization within a given volume. The chamber was fixed to a float which could be anchored

PLATE VII

(*a*) Professor Regener and his apparatus for measuring the intensity of the cosmic rays near the bottom of Lake Constance. (E. Regener.)

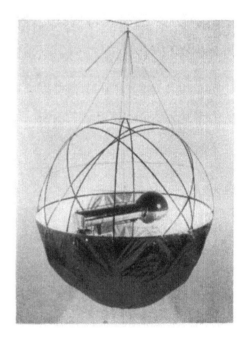

(*b*) The balloons and cellophane chamber containing the ionization vessel for measuring cosmic ray intensities. Professor Regener has sent such an apparatus up to a height of 17 miles.

(*c*) The ionization chamber and registering apparatus for high balloon flights. (E. Regener.)

at various depths, according to the arrangement illustrated by Fig. 19.

In order to remove the possibility that the radioactivity of the water near the bottom differed from that at the top, the chamber was surrounded by a second chamber 2·5 metres in diameter, and filled with water from the surface. The measurements with the protected chamber were very slightly different from those with the unprotected chamber, so Regener concluded his observations were not distorted by radioactivity in the water of the lake. He considered that the analysis of his observations showed the

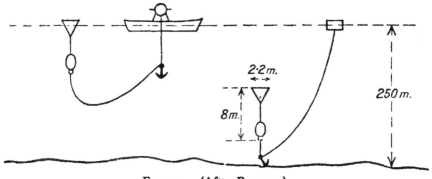

FIG. 19. (After Regener.)

presence of four components in the rays, the relative intensities of the three hardest or most penetrating being about 1 : 6 : 16.

He confirmed his ionization chamber observations by electron-counter tube observations. The cosmic rays may be detected by methods other than those depending on the discharge of electrified filaments through the ionization of the surrounding gas. As early as 1916 Hess and Lawson devised an electrical counting apparatus for counting the number of very penetrating rays passing through a chamber. If the average ray produces a constant amount of ionization, relative intensities of the rays may be compared by counting the number passing through an instrument

in a given time, without measuring how much ionization is produced. This method did not become
practicable until Geiger and Müller introduced their
electron-counting tube. This apparatus consists of a
wire fixed along the axis of a tube. The wire is kept
at a high potential, 1,000 volts or more. When a ray
enters the tube and releases some electrons by ionization they are accelerated by the electric field and
free more electrons by collision. This enables a
sensible current to pass between the walls and the
charged wire, and magnification of the current by
thermionic valves allows it to be registered as a click
in a loud-speaker, or by a revolving counter. The
Geiger-Müller tube will count automatically the
number of rays passing through it.

When Regener sank his chamber containing an
electron-counter tube in Lake Constance he obtained
the following records:

(After Regener: *Nature*, 1931)

Depth in metres.	Total period of three registrations.	Average number of impulses per hour.*
0	1 hour	7,920
1	$\frac{1}{2}$,,	5,500
3	$\frac{1}{2}$,,	4,840
7	$\frac{1}{2}$,,	3,610
9	1 ,,	3,350
18	$\frac{1}{2}$,,	2,000
34	3 ,,	867
93	$4\frac{1}{2}$,,	172·5
133	4 ,,	88·5
183	$5\frac{1}{2}$,,	52·5
235	28 ,,	13

* The residual impulses due to the apparatus (about 500 per hour)
have been subtracted.

The electron-counter tube observations agree with the ionization chamber observations. Kolhörster has announced that he has detected cosmic rays in a salt-mine at a depth equivalent to 800 metres of water. These rays would be four times harder than the most penetrating rays observed by Regener and of at least 100,000,000,000 volt energy. He remarks that these rays would be able to maintain the electric charge of the earth, a phenomenon hitherto not satisfactorily explained.

Besides following the cosmic rays to great depths Regener has followed them to great heights. A self-recording electrometer of the type used in the Lake Constance observations was carried up by a pair of rubber balloons. The position of the electrometer wire was photographed every 4 minutes. The temperature and pressure were measured simultaneously by arranging that the photographed shadow of the wire was cut on one side by the expansion of an aneroid barometer, and on the other by a bimetallic strip that bent with change of temperature. Thus the lengths of the shadow above and below the centre line gave the pressures and temperatures. The apparatus was protected from the low temperature at great altitudes by encasing it in Cellophane material. This trapped heat from the sun's rays, on the greenhouse principle. Consequently the temperature never fell outside the range $+ 15°$ and $+ 37°$ C. at great heights. Thus the possibility of error due to very large temperature variations was eliminated. The apparatus was carried to a height at which the air pressure was 22 millimetres, or one thirty-fifth the normal pressure at sea-level. Thus the apparatus had penetrated thirty-four thirty-fifths of the material of the atmosphere. A pressure of 22 millimetres corresponds to a height of about 27 kilometres or

15 miles. Regener found that the rate of increase in the intensity of the rays decreased rapidly above about 12 kilometres, so that an estimate of the intensity on entering the atmosphere could be made. He found the intensity is equal to that which would produce 333 pairs of ions per second in air at 0° C. and normal pressure, so the intensity of the rays in space is more than 250 times that at the earth's surface. The total number of pairs of ions per second produced in the complete column of air on 1 square centimetre of the earth's surface is 102,000,000. This shows that the cosmic rays have a large energy, about equal to that of star light. They would raise the temperature of a cold body in space to about 3° above absolute zero. Regener's experiments gave the limits of the penetration and intensity measurements recorded before the end of 1932. In 1933 his apparatus was taken slightly higher.

The intensity measurements do not necessarily give any information concerning the direction of the path of the cosmic rays. Myssowsky and Tuwim in 1926 first noticed the relation between the barometric pressure and intensity. Their observations by the Neva in Leningrad showed that the rays are generally weaker when the barometer is high than when it is low. When the barometer is high, the amount of material in the air above it is greater than usual because the pressure is higher than usual. The extra material reduces the intensity of the rays penetrating it.

The direction of the rays was shown by Kolhörster and von Salis during their glacier observations in 1923 to be chiefly vertical at sea-level, and within a cone of 50°. Myssowsky and Tuwim investigated the direction of the rays through the water-tower of the Polytechnic at Leningrad in 1926, and concluded they arrived from the celestial globe uniformly in all directions and the preference to the vertical at sea-level was due to absorption of rays inclined to the vertical and hence having a longer path through the air. In 1931 Barnothy and Forro applied electron tube counters to the determination of direction and confirmed its connection with the vertical. The principle of the important tube-counter method, first used by Bothe and Kolhörster in cosmic ray research, is simple in theory.

Suppose two counters are placed one vertically above the other, as in Fig. 20. A vertical cosmic ray which passes through one will, if it is a particle,

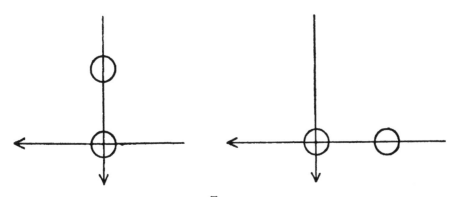

Fig. 20.

generally pass through both, whereas a horizontal ray will pass through one only. If the two tubes are on the same level a horizontal ray which passes through one will generally pass through both. Bothe and Kolhörster in 1929 arranged that each impulse from both tubes should be registered on the same moving film, so simultaneous impulses were at once evident. They found 20 per cent of the impulses were simultaneous, and interpreted their result as showing that the cosmic rays consisted of particles and not rays. The rays were not photons of very high energy, not similar to X-rays but to swift electrons. A wave-ray ends on the atom it ionizes. It communicates its energy to the electron it strikes out of the atom. Hence a wave-ray does not proceed any farther after it has caused an ionization. A particle-ray may strike and ionize many atoms along its path. If it has sufficient energy it may have a long path. Thus two counters cannot be discharged by one wave-ray but they can be discharged by one particle-ray. Bothe and Kolhörster's deduction of corpuscular rays was not conclusive from these experiments alone, because the wave-rays might have struck the electrons very violently and given them an enormous speed. In that case the counters would have been detecting not the cosmic rays, but secondary corpuscular rays produced by the incident cosmic rays. Tuwim found the counters were discharged 20 per cent more frequently by the vertical than by the horizontal rays. Rossi arranged that the counters automatically registered simultaneous impulses only. He found the vertical rays were more penetrating than the horizontal rays.

The introduction of the Geiger counter brought swift progress. The assumption that the cosmic rays were waves, which had been held for twenty-six years, was thoroughly upset. When the presence of very swift particles was demonstrated by the counters earlier evidence of the existence of their tracks was confidently accepted. The tracks were in fact first recognized in 1927. The importance of the invention of new instruments and technique is seen again. The progress of science depends equally on experimental and engineering technique, and on theoretical technique.

The swift particles should leave cloud-tracks in a Wilson expansion chamber. Many of them had probably been photographed by experimenters with Wilson apparatuses, but they had not had peculiarities sufficient

PLATE VIII

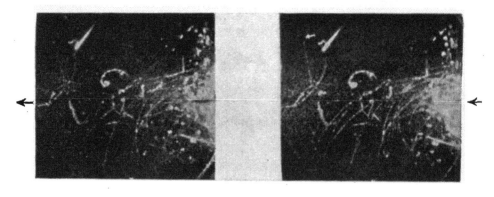

(a)

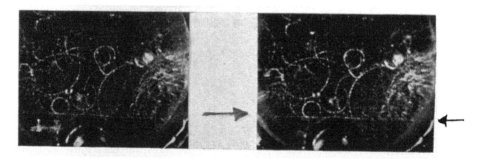

(b)

(a) The first published photograph of a track of a particle connected with cosmic rays. The track is indicated by a small white arrow in the top left-hand corner in the picture on the right. The photograph was taken by Dr. D. Skobelzyn of Leningrad. (D. Skobelzyn : *Zeit. für Phys.*, 1927.)

(b) A pair of stereoscopic photographs of a swift particle connected with cosmic rays, also made by Dr. Skobelzyn. (D. Skobelzyn.)

(c) Dr. Kunze's apparatus for submitting cosmic-ray particles to very powerful magnetic fields. The magnet takes a current of 1,000 amps. at 500 volts, and gives a field of 18,000 gauss. It has shown the presence of particles of 10,000,000,000 electron-volts energy in the cosmic rays. (P. Kunze.) (See page 143.)

to attract notice. A swift particle and a slow particle both make straight tracks when undeflected, and the differences between the tracks are restricted to fine characteristics in breadth and tenuousness. Tracks slightly different from those normally produced by particles from radioactive substances had been dismissed as accidental and due to contamination in the instrument, or to experimental maladjustment. The investigators had directed their attention towards the phenomena of atomic disintegration by radioactive particles, and were not specially looking for another type of phenomenon. Things not looked for are often missed, even when clearly there.

Photographs of the tracks of undeflected particles could not easily lead to the discovery of very swift particles not connected with radioactive substances. If particles are deflected by a magnetic field the swifter they travel the less they are bent. In 1926 D. Skobelzyn of Leningrad started a systematic research on the velocity of electrons struck by wave-radiations from radioactive sources. He intended to determine the velocity of the struck electrons by deflecting them in a magnetic field of known strength. When he began to take photographs he found that while nearly all of the tracks of the electrons were severely bent, sometimes in circles, a few tracks remained almost straight. The visual difference between the two sorts of tracks was most pronounced and indicated the presence of two distinct groups of electrons with distinct orders of velocity. A field of 1,500 gauss left the swift particles virtually undeflected.

Skobelzyn suggested the tracks might possibly be due to the 'runaway' electrons which, as Wilson suggested, must be produced by thunderstorms, and noticed that they seemed to begin outside the chamber. Apart from their straightness, indicating exceptional speed, the tracks were characteristic of electrons. These electrons, accordingly, must possess an energy several times greater than that of the most energetic of electrons known to be ejected radioactively. By 1929 Skobelzyn had obtained thirty-two examples of straight tracks in 613 photographs. He found ionization was produced by them at a rate about equal to the rate at which air is ionized by cosmic rays. The tracks were usually bent towards the horizontal, and the distribution of directions agreed qualitatively with the results of the researches of Myssowsky and Tuwim, and Steinke. The measurement of the degree of deflection showed the presence of electrons of 15,000,000 volts energy, twenty-seven photographs of the swift electron tracks showed double tracks, and one a triple track. The independent appearance of the pairs and the triplets is extremely improbable, so they presumably came from one centre, and such a centre could not be a simple radioactive nucleus. Rutherford had suggested in 1929 that the very penetrating rays, or the electrons struck by them, might strike and disintegrate atoms. Hence these multiple tracks appeared to be the traces of the particles ejected from atomic nuclei by cosmic rays or rays associated with them. The photographing of swift electron tracks was continued by Skobelzyn himself, Mott-Smith and Locher, Millikan and Anderson and others. Mott-Smith found tracks of particles of energy 2,000,000,000 electron-volts. Were these tracks of cosmic rays themselves, or of the secondary particles motivated by them? Millikan and Anderson found tracks of electrons and protons in equal numbers. In a note published in 1932 Anderson commented on a number of tracks which appeared to be due to particles with a positive charge. In one experiment he had placed a lead plate across the Wilson chamber and the particle had passed through it, but with a loss of energy because its track was more bent after emergence, as shown diagrammatically in Fig. 21.

The increase in bending was a proof that the particle had come downwards, and the direction of bending, which depends on the charge, was the contrary of that of known electrons with negative charges. The quality and deflection of the track suggested, Anderson wrote, it was due to a particle of positive

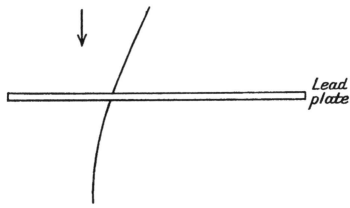

Lead plate

FIG. 21.

charge but of mass less than that of the proton. Anderson did not publish his photographs and did not at the time go beyond the suggestion of the existence of a particle of positive charge and mass less than that of the electron, perhaps a positive electron as a counterpart of the long-known negative electron.

The investigation of the swift particles connected with cosmic rays has also been conducted by Kunze with extremely powerful magnetic fields. The strength of the magnetic field is limited by the size of the Wilson chamber. If iron pole pieces are used to concentrate the field, the chamber must be between them and in a position conveniently photographed. It is difficult to obtain a field more than a few thousand gauss in strength under these conditions. More powerful fields must be obtained by placing the chamber within a wire coil through which a strong current can be sent for a short time. The current is cut off before the apparatus becomes too hot. Kunze arranged with the authorities of the Central Electric Power Station at Rostock to have the use of large currents early in the morning before the load on the station becomes heavy. He sent a current of 500 volts and 1,000 amperes through the solenoid surrounding the chamber, and erected the apparatus within the power station because such large currents could not be taken from the town mains. Two of the mercury rectifiers were switched out of the town mains circuits on to the water-cooled solenoid, giving a current of 1,000 amperes at 500 volts. The solenoid contained one ton of copper wire, and its temperature rose at the rate of one degree per second, so it could not be operated for more than 50 seconds. The current was switched on and off once for each Wilson chamber expansion. The magnetic field was 18,000 gauss in strength (36,000 times the earth's magnetic field), so strong that in the first experiments it prevented the camera shutter from closing, blew out arc-lamps, disturbed measuring instruments and made nails and pieces of metal fly through the air. The current was run for 2 seconds, and the chamber expansions during 3 seconds. Only eighteen experiments could be made during one day because the coil needed 24 hours to cool down. The limits of the field-strength are dependent on the circumstance that it varies as the square root of the current strength. Thus an increase of the current strength by nine times produces a rise of magnetic field strength by three times only.

The photographs of the tracks proved to be well-defined. The lines were thin and susceptible of accurate measurement. Thus a slight curvature could be detected and tracks of radius 5 metres could be measured accurately. A few centimetres of the circumference of a circle of 5 metres radius is not easily distinguished from a straight line when it is made of a streak of cloud.

Kunze found positively and negatively charged particles occurred about equally in ninety photographs showing tracks. The energies of the particles varied continuously, there was no favoured energy value. One of the fastest of the measured particles had an energy of 2,660,000,000 volts, and two others had energies of more than 3,500 and 9,200 million volts respectively.

The vastness of these energies is shown by comparison with the calculated energies of formation of various atomic nuclei. A helium nucleus requires 27 million volts, silicon 216, oxygen 116 and iron 500 million volts. The energy of cosmic ray particles or particles associated with cosmic rays is greater than that of the formation of atoms out of simpler units.

These particles gave almost straight tracks, whereas electrons of 300,000–500,000 volts energy ejected by radiations from a radioactive source were whirled round in small spirals.

The problem of the nature of the cosmic rays was made much more exciting by the definite discovery of very swift particles connected with them. They had been assumed not to be particles because such very swift particles had not been found in connection with any other phenomenon. Kunze's 10,000,000,000 volt particle would have a velocity only 5½ metres less than the velocity of light (300,000,000 metres per second). Also, particles of even 10,000,000,000 volts energy should be sensibly deflected by the earth's magnetic field. Though the magnetic field of the earth is of low intensity its extent enables it to exert a sensible effect on a particle during a passage of hundreds of kilometres. In experiments such as Kunze's a magnetic field of 18,000 gauss bent a swift particle into an arc a few centimetres or inches long. The curvature of the arc was not measurable if it was less than that of a circle 5 metres in diameter. Kunze's strong field has to exert its effect while the particle travels a few inches, but the earth's weak field, say 50,000 times weaker, is able to exert its effect on a particle approaching the earth during a journey of hundreds of kilometres. Thus the earth may be more effective than any artificial magnet for deflecting particles. If the cosmic rays are particles they would be expected to be less intense in regions near the equator than in regions near the magnetic poles. Nearly all of the early measurements of cosmic ray intensity were done in latitudes distant or fairly distant from the equator in North and South Europe, and no very significant differences of intensity were recorded. Assuming intensities at the equator were also the same the lack of variation seemed conclusively to prove that the cosmic rays could not be particles but must be waves, and any swift particles connected with them must be secondarily produced by them. Accurate intensity measurements near the equator were first made by the Dutch physicist Clay. He found the intensity near the equator was less than at higher latitudes. The inertia of accepted opinion prevented the complete immediate acceptance of his important discovery. In 1928 Clay found the intensity at Leyden in Holland was 1·1 J or 110 per cent of the normal, while at Bandoeng in Java, which is near the equator, the intensity was 0·76 J or only 76 per cent of the normal. Clay suggested the decrease might be due to the lens shape of the earth's atmosphere, which is thicker at the equator. Rays coming through would be decreased in intensity.

The problem of the geographical distribution of the cosmic rays inspired observations in every quarter of the globe. Behounek measured the intensity at Spitzbergen. Malmgren and Behounek measured the intensity in an airship over the North Pole. Grant made observations in the Antarctic. All of these observers noted no significant connection with latitude.

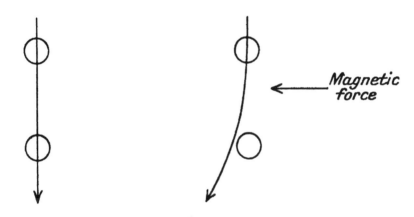

FIG. 22.

But Clay's results, and the effects of laboratory magnetic fields on cosmic rays, increased the conviction that the earth's magnetic field must affect the rays (though Clay had advanced an alternative explanation of the fall in intensity near the equator).

Bothe and Kolhörster showed that if the space between their vertically-arranged counters was subjected to a magnetic field the number of simultaneous impulses in the two counters decreased. Curtiss found a magnetic field reduced the number of coincidences by 25 per cent (Fig. 22).

The increasing evidence that the cosmic rays are particles attracted some of the ablest theoreticians to studies of the experimental results. Heisenberg succeeded in explaining some of the most surprising observations. If an ionization chamber has relatively thin walls of iron or lead the amount of ionization is increased when the thickness of the walls is increased. This seems paradoxical, for one would naturally suppose the thicker material would reduce the intensity of the rays and therefore the amount of ionization. The ionization increases as the thickness of the walls is increased up to about one-quarter of an inch. The increase of ionization is partly due to the density of the lead or iron. More ionization is produced in a unit volume of lead than in a unit volume of air. Hence the secondary rays ejected from the lead increase the ionization near the lead. When the metal is thick these are cut off, and cannot make the air inside the chamber appear as if it were itself ionizing more than normally. Curious anomalies particularly studied by Schindler arise when the absorption by layers made of strata of different material, such as lead and iron, is exactly measured. Johnson succeeded in explaining with remarkable exactitude the surprising increase of ionization and many of the anomalies on the assumption that they were caused by cosmic rays consisting of particles whose energy was of the order 1,000,000,000 electron-volts. Heisenberg had also worked on the same problem. When the great founder of the new quantum mechanics appeared as a supporter of the particle theory, one expected that the wave-theory of the cosmic rays had an uncertain future.

A co-ordinated study of the intensity of cosmic rays at representative stations in every part of the earth's surface became necessary to solve the particle-or-wave question. A. H. Compton organized a world-survey with support from the Carnegie Institution. The world was divided into nine regions, and observers with tested apparatus of identical design were to supervise observations in each region. R. D. Bennett measured intensities in Alaska, California and Colorado. E. O. Wollan observed in Chicago, Spitzbergen and Switzerland; D. La Cour in Greenland, A. Carpe on Mount McKinley in Alaska (he was subsequently killed in a crevasse accident); Benade in India, Ceylon, Java and Thibet; S. M. Maude in South Africa; Byrd and Poulter in Antarctica; Ledig at Huancayo in Peru (he was to sail round Cape Horn and Brazil in his return to U.S.A., observing on the route), and A. H. Compton himself arranged parties in Switzerland, Colorado, Hawaii, New Zealand, Australia, Peru, Panama, Mexico, Northern Canada and Northern United States.

The results of this co-operation confirmed Clay's discovery that the intensity is less near the magnetic equator, to which the figures refer. At sea-level at the magnetic equator the intensity is 14 per cent less than at latitude 50° North or South. At altitudes of 2,000 metres (over 6,000 feet) the difference is 22 per cent, and at 4,360 metres (13,000 feet) the difference is 33 per cent. The change in intensity occurs sharply between latitudes 25° and 45° North and South and latitude 34° is the region of most rapid drop.

Compton regarded these results as a conclusive proof that cosmic rays are particles, probably electrons. He remarked the sun might be their source, as Dauvillier suggested. Störmer has explained many properties of the *aurora borealis* on the assumption they are caused by very much slower electrons streaming from space into the earth's magnetic field. The electrons concentrate along the lines of magnetic force leading to the earth's magnetic poles and render the atoms of the attenuated air luminous at high altitudes.

The very swift electrons constitute the cosmic rays. Dauvillier has found evidence that flocculi (vast glowing clouds containing calcium, which usually hover over sunspots) on the sun contain electric fields at a tension of 1,000,000,000 volts. Perhaps the rays consist of electrons shot from the sun by these fields. Certainly, the rays appear to consist of electrons coming from a distance of several hundred miles from the earth. Piccard on his balloon ascents found at great heights that the number of rays passing in each direction was the same.

Hess reported the rays are 0.3 to 0.5 per cent more intense during the day than in the night. This suggested part of the rays might come from the sun.

The range of an electron of a given speed is much less at sea-level than at great heights, because the air is much denser at sea-level and stops the electron more quickly. The track of an electron of a given speed is bent by the earth's magnetic field. As the strength of the field is substantially the same at great heights as at sea-level, the earth's magnetic field is able to bend the track of an electron travelling at a great height more than the track of an electron at sea-level because it has the opportunity of acting on the electron at a great height for a much longer time.

The curvatures of the circles on which each of the tracks lie is the same, but an observer at the great height would theoretically be able to see a much bigger piece or arc of this circle. At a height of fifteen miles the arc is twenty-five times as long as at sea-level.

In the chapter on the *Stars and the Universe* the ideas of Lemaître concerning the evolution of the universe were discussed. Lemaître requires a primeval atom that disintegrates and emits electrons, protons, photons and other forms of particle and wave energy. He is therefore very much interested in the cosmic rays, as they have a fundamental rôle in his theory, and he also requires sources of very swift particles. Lemaître has studied the theory of the distribution of cosmic rays entering the earth's atmosphere. His calculations show the geographical distribution found by Clay, Compton and others may be explained if the cosmic rays consist of electrons or comparable particles of energy about 10,000,000,000 volts coming equally from all directions in space. In the region of maximum change of intensity at about latitude 34° a majority of the electrons and negatively charged particles will come from the east, while the majority of the protons and positively charged particles will come from the west. The observed particles must have come from points high above the atmosphere, and are reasonably identified as cosmic rays.

In 1925 C. T. R. Wilson suggested the very penetrating radiations might be due to very swift electrons produced during thunderstorms. He had found potential differences of 1,000,000,000 volts could exist between a thundercloud and the earth. An electron in such an electric field would acquire energy sufficient to enable it to perform the penetrations observed in absorption experiments. The swift electrons would travel far from the cloud at high speed, and consequently were named 'runaway' electrons by Eddington. Wilson has estimated that at any moment about 1,000 thunderstorms are in progress on the earth's surface. This might enable very swift electrons to be spread fairly evenly throughout the atmosphere. Schonland has made many observations in the neighbourhood of Johannesburg in South Africa where thunderstorms are frequent. In collaboration with Viljoen he has looked for changes in the rate of occurrence of very penetrating rays during thunderstorms. He has used a Geiger-Müller counter shielded with iron, and arranged that the occurrence of discharges due to very penetrating rays and the abrupt changes in the atmospheric electric field due to lightning flashes should be recorded side by side on the same chronograph tape. A statistical analysis of the records showed for some thunderstorms that there was a strong probability that an impulse on the counter would coincide with a lightning flash, and a smaller probability that the frequency of impulses would increase in the few seconds preceding a flash. None of these effects was observed with storms less than 30 kilometres from the apparatus. With nearer storms the number of impulses tended to decrease when the strongly charged cloud was overhead. This confirms earlier observations of Schonland, who had found positively charged clouds had the greater reducing effect. The very penetrating radiations may be 'runaway electrons' which have been shot upwards by the intense electric fields of thunderclouds, and then deflected back by the earth's magnetic field to produce the characteristic very penetrating effects.

The remarkable variety of ideas and technique in the study of very penetrating radiations is manifest. The contributions of C. T. R. Wilson are noteworthy because in theory and experiment they are original. His thunderstorm theory of the origin of the rays is unlike any other, and the application by Skobelzyn of his cloud-expansion chamber to the tracing of the tracks of ultra-swift particles was one of the major advances in the vast research into the nature of very penetrating radiations. The Wilson apparatus is probably the most astonishing instrumental invention recorded in the history of science. The cloud-expansion chamber researches into atomic behaviour initiated by Wilson have been successfully developed by P. M. S. Blackett, who first photographed the disintegration of atoms.   Blackett has ingeniously adapted the Wilson apparatus for convenient photographing of tracks of very penetrating rays.

The air in a Wilson chamber will produce a cloud along a particle track during part of the period of expansion only.   The effective part is about one-twentieth of a second.   As only about 1·5 particles fall on an area of 1 square centimetre in 1 minute, the chance of a particle falling during an expansion is only about one in 800 per square centimetre of effective area.   When photographs are taken at random tracks are therefore found on a small percentage only.   Skobelzyn obtained about one track in ten expansions, and Anderson found only one in fifty of his photographs showed tracks suitable for measurement.   Blackett and his colleague Occhialini have devised an apparatus which performs an expansion and takes a photograph only when a penetrating ray passes through it.   They have designed an apparatus which makes cosmic rays take photographs of themselves, or their own tracks.   The expansion chamber is placed on its side with counters vertically above and below, so a ray passing through both counters will also pass through its illuminated part.   The movement of the piston which makes the expansion is controlled by a circuit which is completed only when the two counters are discharged simultaneously.   Hence the penetrating ray starts the expansion.   By careful design the time from the passage of a ray through the chamber until the completion of the expansion may be reduced to about one-hundredth of a second.   Blackett calculated how far the ions about which the drops of water condense would diffuse.   He found they would move only 0·4 millimetres in oxygen at 1·7 atmospheres pressure in one-hundredth of a second, so the breadth of the track would be about 0·8 millimetres.   This gives sufficient fineness of definition for very accurate measurements. In his original calculations of the rate of diffusion Blackett made a mistake of a factor of four. This happened to be in the right direction, so he obtained definition four times better than he had expected. A water-cooled solenoid is arranged to produce a magnetic field of 3,000 gauss within the chamber.

When the apparatus is made ready, the experimenter has on an average to wait 2 minutes before a ray passes through and operates it. This agrees with the observed rate of coincidences, which is about two per minute. The experimenter has nothing to do after the apparatus is adjusted. He waits in the darkness and the apparatus is absolutely still. Suddenly there is a click and flash, and he knows he has trapped and photographed the track of a particle which has come, perhaps, from a distant part of the universe. He need not run continuously repeated expansions and grope with mechanical flurry for chance trappings of the rays.

Of the first 700 photographs made by Blackett and Occhialini over 500 showed tracks of particles of high energy. Seventy-five per cent of the 500 successful photographs showed single tracks, the majority of which were not appreciably deflected by a magnetic field of 2,000 gauss. This indicates that their energy is greater than 300,000,000 volts, if they are electrons. The remaining 25 per cent of the photographs show single tracks passing through one counter only, or multiple tracks clearly related to secondary processes occurring when the penetrating radiation passes through matter. Some of Blackett's multiple-track photographs were very remarkable. Eighteen of the original hundred or so each exhibited eight tracks of particles of high energy, while four each exhibited more than twenty. In many of these complicated

PLATE IX

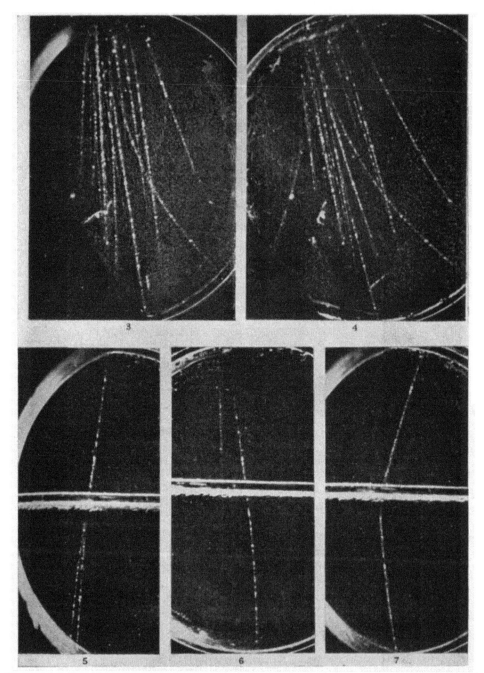

Photographs of positive and negative electron tracks by Prof. P. M. S. Blackett and Dr. G. P. S. Occhialini, taken with the cosmic-ray self-photographing apparatus. Nos. 3 and 4 are a pair of stereoscopic photographs of a shower of particles produced by cosmic rays which have interacted with atomic nuclei. Tracks curved to the right are due to positive electrons or positrons ; those to the left are due to negative electrons. (5) shows a swift particle that has struck a lead plate and knocked an atom forward out of it. The track on the left is of a positron. (6) shows a swift particle that has passed through the plate and is probably a negative electron. (7) shows a track of a positron deflected while passing through a lead plate. It loses speed and its track is consequently more bent when it emerges below. (P. M. S. Blackett and G. P. S. Occhialini and Royal Society.)

pictures numbers of the tracks appear to radiate from points outside or inside the chamber. The tracks are not all deflected in one direction by the magnetic field. Some are bent to one side and others to the other; in fact, many of the tracks seem to appear in pairs, bent and diverging in opposite directions. This indicates the tracks are of particles of opposite electric charges, that some of the particles are negatively charged and some are positively charged and that both sorts appear in about equal numbers. Millikan and Anderson and Kunze had found positively and negatively charged particles appeared in comparable numbers. Kunze assumed the positive particles were protons. Anderson was not sure. His measurements showed some of the positive ray tracks much more than proton tracks. Particles of very high energy produce an amount of cloud along their tracks which is independent of their mass, and dependent only on their charge and speed. Protons and electrons of very high and equal energy produce tracks of very similar quality and appearance, but when they are submitted to a magnetic field the electrons are deflected more than the protons because they are lighter. Anderson concluded some of the positive particles must be much lighter than protons, and reasonably assumed their mass and charge might be equal in size to those of the ordinary negative electron. He advanced this bold view with considerable reserve, and did not immediately publish his photographs with a complete analysis of their features, so his suggestion was not entirely convincing. Blackett and Occhialini's work established the existence of the positive electron beyond doubt. Their large number of photographs and rigid discussion were thoroughly convincing. The study of the cosmic rays had yielded a major scientific discovery.

The showers of particles shown by the multiple tracks may be the débris of atoms disintegrated by the cosmic rays though their nature is as yet obscure. As Rutherford had foreseen, the cosmic rays are proving to be a powerful new tool for investigating the structure of atoms. Their importance is in their high energy. Rutherford developed the modern theory of atomic structure from knowledge gained by the use of the swift helium nuclei ejected from certain radioactive atoms. This was the most powerful of the tools provided by radioactivity. The fast alpha particle has an energy of about 4,000,000 electron-volts. Cosmic rays provide particles of 1,000,000,000 volts, and more. It is not surprising the disintegrations caused by them are much more violent, and may be expected to give new information. The use of cosmic rays may become the most important of all techniques for investigating the structure of the nuclei of atoms, and for a number of years the most fruitful branch of physical research.

The showers of particles from disintegrated atoms usually radiate downwards from some point in the solenoid made of copper wire, and sometimes from points in the glass walls of the chamber, in the aluminium of the piston and the air of the room. When plates of lead or copper are placed across the middle of the chamber radiant points appear in them, too, though none were noted with a tungsten plate. Hence there is evidence that atoms in copper, lead, air and glass may be shattered by cosmic rays.

Somewhat similar tracks might be produced by Wilson's 'runaway' electrons, but these would not have distinguishable radiant points as they would have been shot from a cloud many kilometres distant, and would arrive sensibly parallel.

The shower tracks cannot be completely explained

without the agency of non-ionizing rays such as photons or neutrons. Tracks sometimes originate from
points in the lead plate without the presence of an
incident track, which shows (Fig. 23) that an agent
has reached the plate and violently ejected a particle
without causing ionization during its journey to the
plate. A study of the photographs shows a shower
frequently occurs below another shower. The origin
of very swift photons or neutrons capable of causing
disintegrations offers an interesting problem. They
could not be so confidently expected in the primary

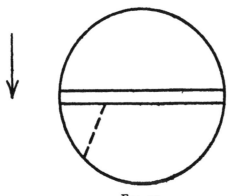

Fig. 23.

beams of cosmic rays owing to the recent proof that
these are probably charged particles.

The experimental discovery of the positive electron
or the positron, as it is now named, confirms one of the
most brilliant theoretical researches in modern physics.
In 1930 Dirac published a theory of the connection
between electrons and protons. These were believed
to be the negative and positive units of electricity.
Dirac tried to deduce the existence of protons, given
the existence of electrons and the laws of quantum
mechanics. He found that the equation for calculating the energy of an electron has negative besides
positive solutions, and he attempted to imagine what
physical interpretation could be given to the notion of

a state of negative energy. He suggested that the
world is almost compactly full of electrons in a state
of negative energy; so full that they cannot be observed.
But a few places, holes, as it were, happen not to be
filled with electrons of negative energy. He showed
that these holes as they moved about the world would
behave like centres of positive energy, in fact like
ordinary electrons. He compared their behaviour
with that of the hole left in an atom when an inner
electron has been removed. As the atom moves
through space it carries its hole with it. The hole
in the atom behaves as a centre of negative energy
because it can be made to disappear only by adding
an ordinary electron of positive energy to it. In
contrast with the hole in the atom, the holes in the
world-compact of electrons of negative energy are
centres of positive energy. Dirac suggested these
holes of positive energy must be protons. Unfor-
tunately, further mathematical investigation proved
such a 'hole' must have a mass equal to that of an
electron. Protons could not be such holes because
they were nearly 2,000 times too massive. The
theory seemed to be only a marvellous fantasy,
though its brilliance suggested the presence of hidden
truth; a fire existed somewhere beneath the flame.
Dirac had deduced the existence of the positive electron, but the complete absence of any other evidence
prevented him from asserting its existence. If he had
declared on the basis of his calculations that the positive
electron existed he would have made the most brilliant
prophecy in the history of science. No one has discovered a major physical entity by deduction only.
Clerk Maxwell discovered the possible existence of
radio waves by deduction, but these are not fundamen-
tally different from light. Max Planck discovered
the quantum of action by deduction, but there was

experimental evidence for its existence. Newton could provide evidence from observational astronomy for the law of gravitation, and Einstein could cite well-known anomalies in the movements of Mercury, swift electrons and photo-electrons in confirmation of his deduced theory of relativity.

Dirac's success with the positive electron has increased the hope that his theory of the possibility of the existence of free unit magnetic poles will be substantiated. He has suggested unit north poles and unit south poles are found together in nature only because the attraction between them, which he has calculated, is great: about 5,000 times that between a proton and an electron. They are not easily separated after they have once come together.

The theory of the positive electron explains why it has been so difficult to discover experimentally. The 'holes' or unoccupied states are easily filled by negative electrons falling into them. Dirac has calculated a 'hole' in material such as water is unoccupied for about $3 \cdot 6 \times 10^{-10}$ second, less than one thousand millionth of a second. Hence the average positive electron never exists freely long enough to be detected in ordinary phenomena. It is seen only when it has such enormous energy that it can make its presence felt within one thousand millionth of a second. The tremendous speed at which it moves allows it to travel about a yard before it is filled up by a negative electron. This is just long enough for it to make a useful track in the Wilson expansion chamber. The marvellous instrument has succeeded in photographing a positron during its life of one thousand millionth of a second, in the progress from its nuclear cradle to its electronic grave. When it combines with the negative electron two quanta of wave energy are formed, two photons. Perhaps these are the non-ionizing agents noted by Blackett, which cause atomic distintegrations.

The mode of production of positrons during the impact of rays on atomic nuclei is the subject of interesting discussion. At present the evidence suggests that positrons are not released from the inside of nuclei during disintegration, but are formed just outside the struck nucleus. A positron is formed as one of a pair of particles by condensation together of a pair of photons impinging on the walls of the nucleus. The other particle is an electron. This theory suggests why positrons and electrons are usually found in about equal numbers.

When the production of positrons during the disintegration of atomic nuclei by cosmic rays was discovered, a search was made for the production of them in other ways. Curie and Joliot had noticed in their neutron experiments that tracks of electrons apparently going backwards were sometimes seen. Chadwick, Blackett and Occhialini exposed a lead target to rays from beryllium, consisting of gamma-rays and neutrons. A number of tracks apparently proceeding from the lead target were shown by magnetic deflection to have a positive charge, and a statistical examination showed they were probably not accidentally produced. By placing a copper plate across the chamber some of these positive particles were proved to be moving away from and not towards the lead, because they lost speed in passing through the copper plate, as the increased magnetic deflection showed. The quality of the tracks and the ionizing power of the particles are similar to those of negative electrons.

The investigation of the cosmic rays has sent men to every region of the earth, to the North Pole and the Equator, to the tops of mountains and beyond into the higher air, and into the depths of mines. Instruments have been sent almost to the top of the atmosphere and to the bottoms of deep lakes. Verigo of Leningrad, for instance, whose results are as yet not generally available, has himself three times carried an electrometer on his back to the top of Mount Elbrus in the Caucasus, and sat there on each occasion for six hours while the cosmic rays registered their intensity. He has made observations in a submarine at a depth of fifty metres, the greatest depth at which direct observations have been made in water, and he has measured the absorbing power of steel by placing electrometers inside the barrels of big naval guns. Refined and ingenious laboratory experiments have been devised, and the great quantum theoreticians have analysed the results. The discovery of a fundamental unit of nature has arisen out of the investigations, and a marvellous theoretical deduction confirmed. Essential information concerning the material constituents of the outer universe has been learned. Is not the study of the cosmic rays one of the most inspiring of human researches?

CHAPTER VI

DIPLOGEN

I

THE vigour of contemporary physical science is in remarkable contrast with the languor of contemporary social life. In a period of contraction of economic activity and of political reaction natural science continues to exhibit within its sphere high creative power and intellectual freedom. The condition of scientific research in America provides a good illustration of this tendency. American research was never more successful. Within a few years several major discoveries have been made, of which the positron is an example, but the greatest American contribution to physical science since the experiment of Michelson and Morley in 1887, which laid the foundation of the theory of relativity, is the discovery of heavy hydrogen, or diplogen. Diplogen is hydrogen consisting of atoms of double the ordinary mass. It is an isotope of hydrogen. 'Isotope' means 'occupying the same place' and refers to the position of an elementary particle in the order of the elements as derived from their chemical properties. Thus all particles with substantially the same chemical properties as hydrogen atoms are named 'isotopes' of hydrogen because they all stand in the same niche in the chemical order of elements. All particles with the same chemical properties as atoms of oxygen are named isotopes of oxygen, and all particles with the same chemical properties as atoms of chlorine are named isotopes of chlorine.

The existence of atoms of different mass and substantially identical chemical properties gradually become evident from the studies of the disintegration of radioactive elements. The number of known elements was rather more than eighty, and yet the researches in radioactivity showed that no less than forty substances with the chemical properties of elements were produced in the disintegrations of the various radioactive atoms. Many of these elementary products of radioactive disintegration consisted of atoms indistinguishable chemically from atoms of previously known elements, yet they had different masses. The number of distintegration products suggested that many of them must be identical or closely related with well-known atoms.

Beautiful evidence for the existence of atoms of various masses and similar chemical properties was given by J. J. Thomson from an entirely different branch of research. He was investigating the nature of the atomic particles which appear in gases bearing electric currents. He submitted the flying particles to magnetic and electric fields which caused them to separate out according to their mass and electric charge. In experiments on the particles in tubes containing the gas neon (now used in the glowing red tubes of advertisement signs) Thomson found that at least two sorts of neon atom appeared to be present as they could be separated by his magnetic and electric fields. One sort of neon atom had a mass of 20 units, and the other 22. The weight of atoms of natural neon determined by the old methods was 20·2. What were the old methods and what did they actually determine? The old chemical and physical methods of determining atomic weights operate with considerable quantities of material. The chemist never works with samples of material containing less than millions of atoms. He arrives at an atomic weight by measuring the number of atoms in a weighed quantity of the elementary substance. The division of the weight by the number of atoms gives the average weight of one atom. The old methods working with large numbers of atoms gives the average weights of atoms. The gas neon as found in the atmosphere contains atoms of average weight 20·2. The ordinary atomic weight methods would give 20·2 even if the gas consisted of a mixture of atoms of mass 20 and 22 which remained thoroughly mixed and in constant proportions.

J. J. Thomson's method dealt with particles singly. Ordinary chemical and physical methods deal only with large groups of particles and average qualities.

The natural science of the nineteenth century assumed a chemical element consisted of atoms of uniform mass on the ground that chemical methods were unable to distinguish between any two atoms. As they were chemically identical they must be identical in mass. The strength of this belief is shown in the last address Kelvin delivered to the British Association for the Advancement of Science, in 1907. In a carefully expressed passage he said:

"It seems, indeed, almost absolutely certain that there are many different kinds of atom each eternally invariable in its own specific quality and that different substances, such as gold, silver, lead, iron, copper, oxygen, nitrogen, hydrogen, consist each of them of atoms of one invariable quality, and that every one of them is incapable of being transmuted into any other."

This strength of conviction is interesting when considered in relation to the evidence even of the old methods of determining atomic weights. The average values of the atomic weights for ordinary substances show a strong tendency towards whole numbers. For instance, the chemical atomic weights of the elements mentioned by Kelvin are: 197·2, 107·9, 207·2, 55·8, 63·6, 16·0, 14·0, 1·008. The atomic weights of all of these elements except copper are distinctly inclined towards whole numbers. A review of the atomic weights of all the elements shows this inclination is marked and therefore probably significant. Some of the more imaginative nineteenth-century scientists were deeply impressed by this and similar regularities. Indeed, Crookes sketched the idea of isotopes in a remarkable address to the British Association in 1886. The tenacious belief expressed so well by Kelvin made the work of the founders of the science of the structure of the atoms of chemistry difficult, as the general opinion of scientists was against the possibility of the existence of such a science.

The technique of analysing rays of particles by magnetic and electric deflections was greatly improved by Aston. In a long series of brilliant experimental researches he showed that the atomic weights of the chemical elements were almost exactly whole numbers, or averages of a few whole numbers. The atomic weight of chlorine proved to be 35·5 because ordinary chlorine consists of a mixture of atoms of mass 35·0 and 37·0, in the proportions 4 : 1. Aston's researches showed that the masses of all atoms could with high exactitude be expressed in terms of the masses of protons and electrons, so that all matter could be regarded as a conglomeration of these two units of electricity. He gave the experimental proof that all matter is made out of electricity.

There were one or two slight discrepancies. The unit atomic weight is that of oxygen and is fixed at 16·00. On this scale the atomic weight of hydrogen is

1·008. The discrepancy can be explained by the theory of relativity and is in fact an indirect proof of the theory. According to the theory energy has mass. Hence consumption of energy is marked by loss of mass. The nuclei of atoms consist of protons and electrons bound together, and energy is consumed in the binding process. Hence the nucleus of the oxygen atom must be less than sixteen times the mass of the hydrogen atom or proton because some mass is lost in holding its constituents together. The discrepancy appeared to be explained very nicely.

In 1929 the American physicists Giauque and Johnston discovered this plausible assumption was incorrect. They were making investigations of the

Two sorts of hydrogen chloride molecule

FIG. 24.

spectra emitted by oxygen. It happens that spectroscopy provides a very delicate method for detecting isotopes under certain conditions. Consider, for example, molecules of hydrogen chloride. This gas when dissolved in water forms the well-known hydrochloric acid. The existence of isotopes of chlorine has been mentioned. Their weights are 35 and 37. The molecule of hydrogen chloride consists of one atom of chlorine to one of hydrogen. Hence the hydrogen chloride molecule may contain chlorine atoms of mass 35 or 37. The total mass in the respective cases will be 36 and 38, and the position of the centres of gravity will be different (Fig. 24).

Owing to the slight difference in balance the two sorts of hydrogen chloride molecule will vibrate slightly differently when disturbed and emit waves of slightly different wave-lengths. The refined methods of spectroscopic measurement allow these wave-lengths to be determined very exactly. Hence the spectroscope can be a very powerful detector of isotopes which happen to provide for comparison suitable pairs of molecules. As ordinary gases such as oxygen normally exist not as free atoms but as molecules containing pairs of atoms, this spinning dumb-bell effect may lead to the detection of isotopes in them. Giauque and Johnston discovered that ordinary oxygen does contain lop-sided molecules. The normal molecule consists of two atoms of mass 16 units; but the spectroscope registered faint lines due to a molecule containing one atom of mass 16 units, and the other of mass 18 units. The difference in the spectra of these two molecules is pronounced because the normal molecule is quite

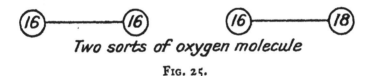

Two sorts of oxygen molecule

FIG. 25.

symmetrical while the abnormal molecule is unsymmetrical. The spectral isotope effect in this case is more pronounced than in the case of hydrogen chloride, because in hydrogen chloride there is only a slight difference between two strongly lop-sided molecules. One is slightly more lop-sided than the other (Fig. 25). Giauque and Johnston also found an oxygen isotope of mass 17 units. By comparing the intensities of the various lines due to the isotopes they deduced that rather less than one thousandth of the oxygen in the atmosphere consisted of atoms of mass 18, and about one ten-thousandth of atoms of mass 17. The existence of isotopes of oxygen was not very surprising but their relatively large quantity was quite unexpected. In 1925 Kirsch and Petterson and Blackett gave evidence from atomic disintegration experiments for the

existence of oxygen atoms of mass 17, but the phenomenon did not seem capable of providing a significant
quantity. Blackett showed by Wilson chamber photographs that atoms of nitrogen from which a proton
was ejected by an impinging helium nucleus sometimes
seized and retained the nucleus. Hence the nitrogen
atom of mass 14 lost a proton of mass 1 and gained a
helium nucleus of mass 4, so that a particle of mass 17
was produced. As this particle had an electrical
charge equal to that of the ordinary oxygen atomic
nucleus its chemical properties would be the same as
those of oxygen and it would be an isotope of oxygen.
The rate at which nitrogen atoms in the atmosphere
are being disintegrated and forming oxygen atoms of
mass 17 is presumably very slow, and the amount of
oxygen of atomic mass 17 produced by the disintegration of nitrogen is presumably insignificant.

The discovery of fairly considerable quantities of
oxygen atoms of mass 17 was not very surprising, but
the discovery that one out of every 1,250 oxygen
atoms in the atmosphere has the noticeably larger mass
of 18 units was entirely unexpected and astonishing.
Evidently atmospheric oxygen was far from being a
simple gas of uniform atoms of mass 16, and the average
mass of its atoms must be considerably greater than
16. The relativity explanation why hydrogen atoms
might have mass a 1·008 if oxygen atoms are 16 must
probably be incorrect or the average atomic weight of
hydrogen atoms must be more than 1·008, because the
average weight of oxygen atoms is in fact greater than
16. This suggested the existence of hydrogen atoms
of mass greater than 1·008. The American spectro-
scopists, Birge and Menzel, calculated that if ordinary
hydrogen contained some atoms of mass 2 a proportion of 1 in 4,500 would be sufficient to make the
average hydrogen atom of the extra average weight

required for the preservation of the accuracy of the relativity explanation.

Isotopes of hydrogen may be expected to have exceptional properties. The ordinary hydrogen atom of mass 1 is very light, and the addition of an extra unit doubles the mass, i.e., makes a difference of 100 per cent in the mass. Ordinary oxygen atoms are of mass 16, so the addition of a unit mass produces an atom of mass 17, which is an increase in mass of only 7 per cent. The chemical and physical changes due to a 7 per cent change of mass in an atom are not easily detectable, but a 100 per cent change may produce noticeable effects. The chemical properties of an atom are almost entirely due to the size of the electric charge on its nucleus. If the electric charge is 8 units of electricity the chemical properties remain virtually the same whether the mass is of 16, 17 or 18 units. The slight differences of mass between the oxygen isotopes allows them to be detected by a delicate technique but makes them extremely difficult to separate. The large difference in mass between hydrogen isotopes might allow them to be separated without very great difficulty.

In 1931, F. G. Brickwedde, of the Bureau of Standards at Washington, and H. C. Urey and G. M. Murphy, of Columbia University, began an attempt to discover possible hydrogen isotopes.

The differences in mass between the atoms of hydrogen isotopes should cause distinct differences in the rate of evaporation from liquid hydrogen. The atoms of mass 1 should evaporate more quickly than atoms of mass 2. Brickwedde allowed a large quantity of liquid hydrogen to boil until a small quantity of liquid was left. The liquid residue was allowed to evaporate separately and the gas from it was collected. In one experiment 6 litres of hydrogen was allowed to boil at atmospheric pressure, and in another 4 litres were allowed to boil at the particular very low temperature at which hydrogen may be liquefied by pressure. The specimens of hydrogen gas collected from the residues left by these two evaporations were examined spectroscopically by Urey and Murphy. The photographic plates must be exposed for a certain period before the lines due to ordinary hydrogen become clearly photographed. Urey and Murphy exposed the plates for a period 4,500 times as long as this, on the assumption that the lines due to the hydrogen isotope would be 4,500 times fainter, according to the proportions of the isotope calculated by Birge and Menzel to be present in ordinary hydrogen. Faint but definite lines due to the isotope appeared on the photographic plates. Comparisons of their faintness indicated that the residual gas from the first evaporation experiment contained one isotope atom to every 4,000 ordinary atoms, while the gas from the second experiment contained one in 800.

This brilliant discovery stimulated the search for other methods of separating the two hydrogen isotopes. Washburn and Urey attempted separation by electrolysis. When two electrolytes such as solutions for potassium hydroxide and sodium hydroxide are mixed and electrolyzed the substance which is decomposed at the lower voltage is decomposed first. Hence the molecules of electrolytes containing one of the isotopes should be decomposed more quickly than the molecules containing the other isotope. This effect may be partly due to the difference in the mobility of the two isotope atoms. An atom of mass 1 may be impelled through the electrolytic solution more quickly than an atom of mass 2 because it is lighter.

An attempt to discover the hydrogen isotope by electrolysis of water was begun before the isotope had been discovered by evaporation of liquid hydrogen. When water is decomposed by an electric current the light hydrogen atoms should pass through the liquid more quickly and the collected hydrogen gas should have an excess of light hydrogen atoms, while the remaining liquid should have an excess of heavy hydrogen atoms. Besides starting experiments on the electrolysis of water, Washburn and Urey started an examination of the liquids in the electrolytic cells of commercial electrolytic factories, where metal-plating and chemical manufacture by electrolysis are done on a large scale. Often the liquids from the cells are not changed for several years; water being added merely to repair the loss through evaporation, as it is added to the acid in the cells, or liquor as it is named, might contain a concentration of heavy hydrogen atoms through the continuous operation of the electrolytic separating effect. Liquor from the cells in the plant of the Southern Oxygen Company of Virginia, which had been used continuously for two years, was examined, and later, liquor used for three years from the Ohio Chemical Company of New York. Urey, Brickwedde and Murphy showed these residual solutions contained an excess of heavy hydrogen atoms.

Water consisting of molecules of oxygen and heavy hydrogen atoms must be denser than ordinary water. The increase in density will not be great, as the mass of the ordinary water molecule is 18 units, while with the heavy hydrogen the mass would be 19 units.

Washburn, Smith and Fransden found the specific gravity of water increases as the electrolysis proceeds, the temperatures of freezing and boiling rise, and the refractive index for light decreases.

They found also that water made by combining oxygen obtained from ordinary water by electrolysis with an excess of ordinary hydrogen has an abnormally low specific gravity. If this artificial water is electrolyzed and its oxygen combined with excess of normal hydrogen, the second lot of manufactured water has a still lower specific gravity. Evidently the oxygen picks out the lighter hydrogen isotope from the excess of hydrogen atoms. The lighter hydrogen atoms are more mobile and presumably would react more quickly.

These authors are engaged in a survey of the water obtained from different natural sources, in order to find whether the hydrogen isotopes have important differences of rôle in the processes of animate and inanimate nature.

The eminent American physical chemist, G. N. Lewis, has energetically followed the investigations of the hydrogen isotope. He had been interested in finding methods of separating the oxygen isotopes, but this was necessarily a very difficult research because the slight differences between the oxygen isotopes provide a very small basis for the operation of separating techniques. He turned to the much more promising field of research offered by the hydrogen isotopes. In his researches on the oxygen isotopes he had measured the specific gravity of specimens of water by finding the temperatures at which a 10 c.c. float will neither sink nor rise. The comparison of the temperatures gives a comparison of the specific gravities. When the samples of water were distilled and freed from dissolved air he could measure their specific gravities to an accuracy of one part in 1,000,000.

In his laboratory Lewis happened to have an electrolytic cell containing liquor four years old. The specific gravity of distilled samples of this liquor compared with that of normal water was 1·000034, which indicated a content of hydrogen atoms of mass 2 to an extent of 1 in 3,000. He electrolyzed a quantity of the liquor with a current of fifteen amperes until its volume had decreased by one-third. He found the increase of the specific gravity of the distilled liquor over the normal was 50 per cent greater than in the first sample. This increase might have been due to a concentration of heavy oxygen isotopes. In order to test the possibility Lewis distilled some of the electrolyzed residue and passed the steam over iron. The oxygen was removed from the steam by the iron, and the free hydrogen continued on through a tube containing hot copper oxide. The free hydrogen combined with oxygen in the oxide and formed steam again, which was condensed into water. Then a stream of ordinary hydrogen was passed over the hot iron oxide, so that steam and hence water was again produced. This second specimen of water made from ordinary hydrogen and presumably ordinary oxygen had the normal specific gravity, but the first specimen of water made from the hydrogen obtained from the liquor had the same specific gravity as the liquor. Hence the abnormal specific gravity of the liquor water was due to abnormal hydrogen and not to possibly abnormal oxygen.

Lewis then started an experiment on a larger scale. He arranged to electrolyze 10 litres of liquor from the old electrolytic cell down to a volume of 1 cubic centimetre. The electrolysis was done in a vessel that could be cooled by a copper coil through which cold water could be circulated. Soda was dissolved in the heavy water to make it conduct electricity and pass a current of 250 amperes. In the later stages of the decomposition the strength of the current was reduced. After five days the original 10 litres had been reduced to 1 litre. Most of this litre of residual liquor was placed in a vessel surrounded by cold water and carbon dioxide gas was bubbled through the residual liquor in order to neutralize the soda in it. This solution was heated in a copper vessel and the pure water was collected in a condenser, giving a volume equal to that taken from the residual liquor. This volume of purified water was then mixed with the remains of the residual liquor from which the soda had not been removed, making up the volume of liquid to the original litre. The prepared litre was electrolyzed down to a volume of 100 c.c., and by similar methods the 100 c.c. was electrolyzed down to 10 c.c., and the 10 c.c. down to $\frac{1}{2}$ c.c., i.e., to a volume one twenty-thousandth of the original 10 litres of liquor. This half of a cubic centimetre of liquid was treated with carbon dioxide to remove its soda, and distilled. The pure water had a specific gravity of 1·035, and therefore 31·5 per cent of its hydrogen atoms consisted of the heavy sort.

In his next experiments Lewis electrolyzed 20 litres of liquor with a current of 400 amperes in a cylindrical vessel made of monel metal. The 20 litres were reduced to $\frac{1}{2}$ a cubic centimetre of a liquor which gave water of specific gravity 1·073 and hence contained among its hydrogen atoms 65·7 per cent of the heavy isotope.

Lewis deduced that under equivalent conditions the light hydrogen atom is evolved five times as quickly as the heavy atom in electrolysis. If water made of hydrogen containing 65·7 per cent of heavy atoms were reduced to one-quarter of its volume by electrolysis, the residue should consist of a water whose hydrogen would be 99 per cent heavy isotope.

In conclusion, Lewis deduced that one out of every 6,500 hydrogen atoms in ordinary water is a heavy isotope atom of mass 2.

Lewis's estimate of the comparative rates of evolution of the light and heavy isotopes was as $5 : 1$.   J. D. Bernal and R. H. Fowler have shown by calculations based on quantum mechanics that the light hydrogen atom is probably five times as mobile as the heavy hydrogen atom in water.

In still later experiments Lewis prepared water $99 \cdot 99$ per cent of the hydrogen of which consisted of the heavy isotope.   At $25°$ C. this very pure heavy water has a specific gravity of $1 \cdot 1056$.   It freezes at plus $3 \cdot 8°$ C. and boils at $101 \cdot 42°$ C.   The maximum density of ordinary water is not at $0°$ C. but at $4°$ C. Heavy water shows a parallel abnormality, but its density is at a maximum at $11 \cdot 6°$ C.   In the qualities which make ordinary water an abnormal liquid heavy water is still more abnormal, but the degree of abnormality decreases with rising temperature.   It has been announced that Herz has obtained diplogen so pure that its spectrum failed to show the presence of ordinary hydrogen.   Herz applied his serial diffusion apparatus to accomplish the separation of the two isotopes.

The differences between ordinary water and heavy water are evidently considerable and the problem of their effects in nature is of great interest.   Lewis has written that he was interested from the beginning in the possible biological effects of heavy water, and has already started some biological experiments with it. He has experimented with the seeds of the tobacco plant (*Nicotiana tabacum var. purpurea*) to discover whether they will germinate in heavy water.   In ordinary water and ordinary conditions they germinate very reliably.   He put pairs of seeds in six tubes, three of which contained some ordinary water and three some heavy water.   The tubes were sealed and placed in a thermostat which preserved a constant temperature of $25°$ C.   Within two days the seeds in the ordinary water sprouted and at the end of two weeks formed well-developed seedlings. The three pairs of seeds in heavy water showed no development. H. S. Taylor, W. W. Swingle, H. Eyring and A. H. Frost state that tadpoles of the green frog (*Rana clamitans*) could not live in it for more than an hour. The single-celled protozoan (*Paramoecium caudatum*) lived in it for two days only. The common aquarium fish (*Lebistes reticulatis*) is killed in two hours. The flatworm (*Planaria maculata*) is killed in three hours.

This difference in behaviour of living organisms with the two sorts of water suggests that living organisms may be able to separate isotopes.

Recent research has shown that at least three sorts of oxygen and two of hydrogen exist, hence there must be nine possible sorts of water. The respective rôles of these in nature remain to be investigated, and their determination may reveal new regions of knowledge. The chief constituents of living matter are oxygen, hydrogen and carbon. The known number of compounds of these substances, the subject-matter of organic chemistry, is about one million. The discovery of three sorts of oxygen, two of hydrogen and two of carbon adds to the variety of the possible compounds of organic chemistry by millions. The discovery of heavy hydrogen has presented the organic chemists with generations of research. Besides variety in oxygen and hydrogen composition water exhibits other peculiarities. Bernal has developed a theory that water has three sorts of structure. The first type of structure which is named tridymite is shown by water when, in spite of liquidity, its molecules tend to arrange themselves in the same order as the atoms in an ice crystal. This state is rather rare and occurs at temperatures below 40° C. The second type of structure is predominant at ordinary temperatures. The molecules in the water tend to arrange themselves in a formation similar to that of a quartz crystal. In the third type of structure the molecules are packed closely, as in an ideal liquid such as liquid ammonia. It is predominant at high temperatures below 374° C.

Recent research shows that water is not simple, but complicated in its composition and properties.

The heavy isotope of mass 2 has already been used in physical experiments. Professor E. O. Lawrence has employed it as a projectile in atomic disintegration experiments. As hydrogen nuclei of mass 2 are twice as heavy as ordinary protons, they might be more effective disintegrators in experiments such as those made by Cockcroft and Walton and already described. Lawrence bombarded atoms of gold and platinum with heavy hydrogen nuclei, or diplons, as they have been named, moving under an electric field of 1,500,000 volts. He found the diplons did not disintegrate the gold and platinum atoms but were themselves disintegrated, with an emission of 7,500,000 electron-volts of energy. The energy released in each disintegration was much greater than the energy of the impinging particle. Lawrence believes the diplon splits into a proton and a neutron. He found an increase of half a million volts in the energy of the diplon produced an increase of only a quarter of a million in protons from diplon disintegrations. This suggests the energy is being shared between particles of equal mass. The proton and the neutron are of almost exactly equal mass. He found that an atom of carbon when struck by a diplon retains the neutron part but not the proton. An atom of higher atomic weight is synthesized out of atoms of lower atomic weight, as the oxygen isotope of mass 17 is synthesized when nitrogen nuclei are struck by helium nuclei.

The production of neutrons from the nuclei of the heavy hydrogen isotope obtained from water shows that neutrons may be obtained from water. Rutherford and Oliphant have shown diplons disintegrate according to rules different from those for protons. Bainbridge has shown by a special mass-spectrograph that the mass of the diplon is 2·0136 units.

These discoveries are a gratifying demonstration of the vigour of contemporary science. Scientists never found themselves surrounded with more promising problems. The discovery of the laws of so large a variety of new phenomena must lead to the invention of new powers of control over nature.

Urey, Murphy, and Brickwedde have suggested the ordinary hydrogen atom should be named 'protium' and its nucleus the 'proton', while the heavy hydrogen atom should be named 'deuterium', and Lewis suggested its nucleus should be named 'deuton'. Rutherford and others have suggested the new heavy hydrogen should be named 'diplogen' and its nucleus the 'diplon', because these names are easily distinguished from all others besides being accurate.

The naming of new atomic particles is an agreeable luxury, of which the present scientific period is enjoying an exceptionally large share. The discovery of diplogen has been a great triumph for the American schools of spectroscopy and physical chemistry. In its sphere it is as important as the theory of relativity in another sphere. Hydrogen is the most reactive element and enters into more compounds than any other. Hence the variety of substances containing hydrogen is at least doubled. In fact, it is multiplied immensely, for even the relatively simple substance water now appears in nine varieties. Diplogen has presented organic chemists with generations of new research. Besides providing new varieties, it should be a powerful probe of organic compounds. For instance, it is found that hydrogen and diplogen behave differently when in similar places in a molecule of sugar. The substitution of a diplon for a proton will become a standard method of exploring the structure of chemical compounds and discovering the relations between a hydrogen atom and its neighbouring atoms.

CHAPTER VII

THE CHEMISTRY OF EVOLUTION

I

THE theory of the evolution of animals and plants was founded on studies of their shapes. The shapes of animals were not necessarily the first of their qualities to be accurately noted, as man may have learnt tigers are dangerous before he noticed their shape, but his knowledge of animal shapes is very old, as the upper paleolithic cave drawings show. Man could make good line drawings of animals ten thousand years ago. He left pictures of bisons exhibiting remarkable draughtsmanship and colour-sense, and he even left anatomical drawings. A paleolithic drawing of a mammoth containing an indication of the place and shape of the heart is extant (Fig. 26). This is the oldest known picture of the heart of any animal. In pictures of bisons the position of the heart was sometimes indicated by an arrow.

The beginning of the study of animal shapes is coincident with the beginning of a branch of art. Though animal behaviour is more important than animal shape, it did not become a material of artistic and scientific study because it was more intractable. Shapes may be depicted by a relatively simple technique, by lines on a surface. The description of motion and the behaviour of living organisms cannot be accomplished with so simple a technique as lines;

it requires a cinema film exhibiting motion and sound. Even a developed language is not sufficient. Early poetry gave many descriptions of animal and human behaviour in which an unsuitable technique was within its limitations brilliantly employed, but words are less suitable for the description of behaviour than lines for the description of shape. The study of animal shapes progressed because it was relatively easy. In the time of Aristotle a considerable knowledge of the external and internal shapes of organisms

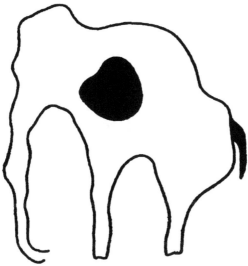

FIG. 26 (after Elliot Smith: *Human History*).   A Paleolithic drawing of a Mammoth showing the heart.

already existed. Medicine and art stimulated the study, as the surgeon must know the structure of a body in order to operate and the artist in order to depict. Anatomy and sculpture have in history often notably developed in the same period. In classical Greece and in the Renaissance the cultural leaders often combined a knowledge of art, anatomy and science. Aristotle and Leonardo da Vinci are examples. When a knowledge of the shapes, the anatomy, of a large variety of organisms existed, comparative studies could be expected to produce evolutionary theories. Leonardo himself compared the shells from the strata on mountains and saw that they resembled those found in the sea. He deduced that the strata must in the past have been beneath the sea, and so founded evolutionary geology.

The comparison of fossil skeletons showed how many species had changed gradually through periods of millions of years. The rows of bones from fossil horses showed a gradual change, for example, in the size and arrangement of the toes. Skeletons of apes could be arranged in a series of which a human skeleton was not an unreasonable end. Darwin studied the contemporaneous distribution of species, how plants and animals of similar yet different species existed at the same time in various parts of the globe. Must not the contemporaneous organisms of similar yet different structure have had common ancestors? The theory of evolution was born of a comparative study of shapes. The idea of a continuous movement of change by which the rows of shapes were connected was furnished by the psychology of the investigator, who unconsciously received it from the process of social development. A contemplation of the process of life as exemplified particularly in the life-history of human beings did not provide a dynamic conception of evolution. The single human life appears to pass through a circle of birth, growth and death which excludes evolutionary advance. The dynamic conception of phenomena could be learned only from entities whose periods of development are much longer than those of human life. Specific forms of social organization seem to require at least several hundred years to pass through their evolutions; if they have a cycle it is much longer than the human cycle of seventy years, and persons living in the middle period can view it as a progressive evolution. The rapid

social changes simultaneous with the growth of production of food and goods in the last four centuries
had created an awareness of the dynamic aspect of
phenomena. The discovery of new continents inspired by the desire for trade and imperial expansion
produced the exploring spirit. Novelty became one
of the characteristics of ordinary life. Men became
accustomed to expect movement in everything. In
the nineteenth century the expression of change
became more defined as the structure of capitalist
society grew more definite. Darwin received a very
significant impulse from Malthus' discussion of the
problem of population. He arrived at the idea of
natural selection from a reading of Malthus' discussion of a social problem. By that time the data
of abstract subjects had become rich.

Dynamic conceptions of phenomena had become
habitual to the human mind. Physical facts had been
conceived dynamically in the Newtonian mechanics.
The desirability of social reform had become respect-
ably accepted. There was a wide belief in the reality
of progress. The last region of human activity per-
meated by dynamic conceptions was that concerned
with man's ideas of himself and living things. The
permeation came through the application of the
conception of movement to the array of data accumulated by biological observation, and the data were
mainly of shape and structure because these are the
qualities of living things most easily studied.

The data for establishing the fact of evolution need
not necessarily be morphological. They may be
behaviouristic. Comparative studies of the behaviour
of apes, such as those of Zuckerman, provide evidence
for consanguinity between different species which
is not morphological. The comparative study of the
sexual life of different species of monkeys shows

relationships between the species different from those deduced from the comparative studies of their anatomies. The line of the evolution of behaviour is possibly the more fundamental. The conventional classification of the lemurs, tarsius and apes depends on the definitions made in 1873 from studies of their anatomy. Their behaviour or physiology provide equally good classificatory data. For instance, they may be defined by breeding season. Thus only the lemurs have a breeding season. Again, lemurs are distinguished from all apes, monkeys and men by their lack of stereoscopic colour vision.

The Old World apes are differentiated from others by sharing with men certain properties that appear in the blood-agglutination test. Lemurs and tarsius do not use hands in grooming, when they pick bits out of the hair. Apes have every variety of facial expression, whereas tarsius have little and their temper is difficult to gauge.

The lower stages in the evolution of intelligence may be studied experimentally, and the results obtained are of greater validity than the inferences based on the studies of the anatomy of the brain. Primates may be graded by the degree to which the higher parts of the brain have taken over the functions of the lower centres. Experiments have shown that the frontal lobe of the brain is not more essential for intelligent behaviour than the parietal; a conclusion which conflicts with many current views but is supported by the fact that the brain of a chimpanzee possesses relatively more frontal lobe than the human. These and other experiments on the brain show that similarities of behaviour may be more important than similarities in shape in determining the relation of animals and the ancestry of man. As improvements of technique provide the possibility, other methods of studying the evolution of living organisms are employed. Living matter consists of a variety of substances, of solids, liquids and gases, which is not infinite in type. For instance, all living bodies contain proteins. While the variety of proteins is immense and each protein may be more or less specific to each species, all proteins have fairly similar properties. The limitation of the types of substances occurring in the constitution of living matter is not surprising because not many types of substance could be expected to possess the peculiar properties of forming matter that lives. A comparative study of the composition of the substances of which living organisms are made should reveal chemical similarities between different sorts of organisms. The occurrence of identical or similar chemical substances in different sorts of organisms may indicate an evolutionary relationship. If a comparative chemical study of the material of living organisms is to provide significant information, it must be capable of refined discrimination. The chemical analysis of the material of living bodies has only recently begun to possess the degree of refinement necessary to provide useful comparative chemical data. Methods of detecting the presence and estimating the quantities of the complex substances occurring in living material are being developed. This successful development of biochemistry has increased the possibility and hence the interest of the study of the chemical relationships between organisms.

During the last twenty years much progress has been made in the chemical study of the colour pigments in flowers. Willstätter has applied a masterly technique of organic chemistry to the elucidation of the nature of these pigments, and his first important success was made by the discovery of the constitution of the colouring matter of the blue cornflower, a substance named cyanidin. The study of the chemistry of plant pigments was widely extended by Robinson and his colleagues. They have shown, from a study of many different orders of plants, that nearly all of the innumerable colours of flowers are derived from three substances only. These are cyanidin, dephinidin and pelargonin. Plants appear to be able to produce the colours of their flowers from compounds of one or another of these substances. The violet colour of the corolla of the fuchsia differs from the outer bluish red of the petals only through the presence of tannin. In ordinary lilac the white flowers, the pale mauve and the deeper red are produced from the same pigment, according to the quantity present and reaction with a modifying substance. All the colours of the rose, the dahlia and the blue cornflowers are probably due to modifications of one pigment; the different colours being due to the combination of the pigment with a variety of modifying substances. The general restriction of the colouring mechanism of flowers to variations in the use of three fundamental pigmentary substances must have important evolutionary significance. It suggests at once that the number of substances having a fundamental rôle in the structure of living organisms is small, and that living organisms are descended from a common parent or a few ancestors that were built out of these peculiar substances in their crudest form. The vast variety of flower colours is an embroidery on the chemistry of a few fundamental pigmentary substances. The chemical constitution of the pigment of a flower is at least as definite a characteristic as the flower's shape. It has equal importance in deciding the relationship of plant orders and species, and the course of evolutionary sequence. It contributes fundamental information concerning the origin and function of flower colour.

II

The development of technique which now enables the chemical characteristics of living matter to be studied profitably has provided the opportunity for the establishment of chemical embryology as an independent branch of science. The task of organizing the disconnected knowledge of the chemistry of embryos was performed by Joseph Needham in his remarkable book *Chemical Embryology*. He made the first summary of all the recorded researches on the chemistry of embryos. Before the publication of this book the field of chemical embryology remained undefined and unordered. There were individual records of chemical aspects of embryos, but no general summary allowing all the facts of the chemistry of embryos to be surveyed simultaneously. Without such a summary comparative chemical embryology could not be pursued effectively, because the various chemical facts could not be compared before they were assembled. Needham mobilized the troops of facts for one of the important scientific campaigns of this generation. The fact of evolution was established by the study of shapes. The mechanism of evolution cannot be deduced from the study of shapes alone because shapes by themselves are in a sense empty; the study of the constituents is also required. The mechanism of a steam engine cannot be fully understood from a study of the external form. An exclusive study of shape, of morphology, is dangerous because the researcher has a longing to fill the empty shapes with a motive power. Morphology is the parent of much vitalistic speculation because the morphologist tends to regard the organism as an animated shape rather than a full, complicated self-acting mechanism. The mechanism of evolution is to be deduced from the constitution of the materials that grow and evolve. The changes during the early stages of growth are the most violent and therefore the most observable. Here the materials of life are working most swiftly and present themselves to the student most vividly. The improved chemical technique allows the subtle chemical changes occurring in growth to be elucidated.

The chemical constitution of the eggs of various sorts of animals has a direct evolutionary significance. When the eggs of simple animals such as sea-urchins are analysed they are found to contain solid organic substances such as protein (fibrous material), carbohydrates (sugary material), fats, etc. They contain only a small quantity of salts, and water. The salt-content is usually described as ash, what is left when the egg is burnt or incinerated. The eggs of invertebrate, non-spinal, animals that live in water do not contain ash. As the embryos in the eggs grow they obtain their necessary supplies of ash or salt from the surrounding water. Hence salty water, such as sea-water, is suitable for the support of these embryos. In the ascent from the lower to the higher animals the eggs show an increasingly complicated chemical content that makes them more and more independent of their environment. Needham has exhibited this relation in a table reproduced in part on the opposite page.

The + sign indicates that the egg contains enough of the substance to make one embryo. The — sign indicates that the egg has not been provided with the substance by its parent organism, and must therefore obtain it from the environment. The o sign indicates that the parent organism has not provided adequate supplies in the egg but provides them in other ways, e.g., by circulation of the blood with its contents from the mother to the embryo in her womb.

The table suggests the increasing independence of the environment possessed by the eggs of parent

THE EVOLUTION OF THE EGG (after Needham)

	Solids.	Ash.	Liquid.	Gas.
	Proteins, fats, carbohydrates, etc.	—	Water.	Oxygen.
Water-animals:				
Eggs with yolk—e.g., Sea-urchins	+	—	—	—
With much yolk, e.g.—				
Sea-gooseberries . . .	+	—	—	—
Octopus	+	—	—	—
Sand-crab.	+	—	—	—
Fish such as trout . .	+	+	—	—
Dog-fish . .	+	+	—	—
Land Animals:				
Frog (amphibian) . . .	+	+	—	—
Turtle	+	+	—	—
Land-frog	+	+	+	—
Snake.	+	+	+	—
Fowl	+	+	+	+
Grebe	+	+	+	+
Man	o	o	o	o

organisms high in the scale of organization. The eggs of the elementary animals are very dependent on their watery environment. They have to obtain all their ash from it, their water, and the oxygen needed in their respiration. This points to a fundamental connection between elementary organisms and this environment. The elementary organisms are suited to growth in salty, aerated water, and this suggests that elementary organisms originated in the sea. The scarcity of invertebrate animals in fresh-water habitats confirms the suggestion. There are no fresh-water urchins and few fresh-water shell-fish and worms. The vast majority of the invertebrate animals have not succeeded in producing eggs adapted to a water deficient in salts. Marine shell-fish have been acclimatized to fresh-water, but no one has succeeded in breeding them in fresh-water. Gurney states that the jack-shrimp of the Norfolk Broads, which lives equally well in fresh-water or sea-water, always goes to the sea to hatch its eggs because they will not develop in fresh-water. The eggs of these animals do not contain enough ash to enable them to develop in fresh-water. A remarkable example of incomplete adaptation is seen in the trout. Its eggs contain ash and develop in fresh-water, but trout spermatoza die much more quickly in fresh-water than in slightly diluted sea-water.

The successful adaptation to fresh-water of animals evolved in the sea may be due to a mutation which causes the eggs to contain ash. The ash-containing eggs could develop in fresh-water and allow its colonization. Ellis remarks that colonization of fresh-water may occur very quickly when the organism has become independent of a handicap such as ashless eggs. The evolutionary emergence of animals from the sea may become possible by a slight change in constitution as effectively as by a change in external or macroscopical structure.

Ephrussi and Rapkine found that the sea-urchin egg obtains ash from sea-water at rates exemplified by the following figures:

Hours after fertilization	.	.	.	0	12	40
Total ash percentage dry weight	.	.	1·5	9·1	16·8
Total ash percentage wet weight	.	.	0·34	2·06	3·56

In contrast with these figures, Hayes found that the ash-content of certain fish-eggs remained constant at 1·5 mgm. per egg throughout development. Hence this fish-egg did not obtain ash from the sea. Herbst demonstrated the absorption of ash by sea-urchin eggs in a series of beautiful researches. He made up a large number of artificial solutions resembling sea-water but without one constituent. He attempted to develop sea-urchin eggs in all of these solutions, and with some he was successful and with others unsuccessful. He found that the sea-urchin egg would not develop normally in solutions lacking phosphorus in the form of calcium phosphate; sulphur in a sulphate; chlorine, sodium, potassium, magnesium; calcium as calcium carbonate or calcium sulphate; and iron. It developed normally in the absence of bromine, iodine and silicon. Herbst found that the egg needed some of the former substances for its own uses and not merely as factors for producing a constant environment. Others were needed for medical protection; for instance, he found that some phosphate was necessary to precipitate traces of poisonous copper in the distilled water of the Naples biological station with which he made his artificial sea-water. The eggs could not flourish until the copper was removed. Then Herbst investigated the interchangeability of the various similar salts. He found sulphate could not be replaced by sulphite, nor by thiosulphate, selenate or tellurate. Chlorine would be partly replaced by bromine, but not by iodine. Potassium could be partly replaced by rubidium or caesium, but not by sodium or lithium. Calcium carbonate could not be replaced by magnesium, strontium or barium carbonates. Sulphur could not be obtained from sulphates of ether when inorganic sulphates were absent. Further research showed that normal development is possible with one combination of salts only, containing sodium, potassium, magnesium, calcium, chlorine, sulphur, carbon dioxide and hydroxyl. Herbst continued his masterly investigation by showing that the salts could be divided into two classes, those necessary from the beginning for development, and those necessary only after a certain stage. The first class included chlorine, hydroxyl ions, sodium, potassium and calcium. The absence of these salts was associated with various defects. For instance, hydroxyl excess was in one case associated with the appearance of pigmentation; ciliary motion (the waving of hairy structures) depended on the presence of potassium; and skeletons could not be formed without calcium. The second class includes sulphate, carbonate and magnesium. Without sulphate the stomach, skeleton and pigmentation would not develop. Carbonate was essential to the skeleton, and magnesium to the stomach and skeleton. Thus Herbst showed the sea-urchin embryo and similar organisms have a very restricted power of utilizing the chemical contents of the environment. They are closely dependent on the chemical constitution of the sea, which suggests the sea was the source of their ancestors, and that they have not progressed far in achieving independence of their source. The primitive organisms are not sturdy independents. The strength of the crude and the primitive is largely illusory. This is probably true of most organizations, and if argument by analogy is permissible it follows that a highly organized social order probably has more power of survival than a less highly organized social order. The present popular demand for a return to primitive forms of government such as personal dictatorship is biologically retrograde, in so far as the biological analogy is applicable. The strong simple man is a contradictory phenomenon, as simple organisms have less power of adaptation and hence less power of control over the environment. The rôle of the water-content of eggs may be considered first from a study of the hen's egg. Satisfactory data exist only of eggs at the fifth day of incubation or later. On the fifth day the embryo consists of 94 per cent of water, and this percentage decreases steadily to about 80 per cent at hatching on the twenty-first day. During this period the embryo is growing, so the weight of water in it is increasing. The loss of weight during the development of hens' eggs has been known since the eighteenth century. Groebbels and Möbert found the same phenomenon in other birds' eggs. They found, for example, that hawks' eggs lost 11·5 per cent, nightingale eggs 11·8 per cent, robin eggs 15·5 per cent, sparrow eggs 16·8 per cent and chaffinch eggs 26·8 per cent. The loss of weight of hens' eggs is so constant that Zunz has suggested that frequent weighing should be used as a guide to normal development. Researches by Murray showed that the loss in weight was chiefly due to the dryness of the surrounding air; in air of 100 per cent humidity the egg lost no weight. There was no difference in the rate of the loss between fertile and infertile eggs until the sixteenth day, when the fertile eggs began to lose weight more rapidly. These results show hens' eggs cannot control the rate at which they lose water, and agree with observations by Aggazzotti, who found that at Col d'Olen, which is 800 feet above sea-level, the eggs lost water more quickly than at Turin, which is on the sea-level. This is contrary to the behaviour of the adult hen, which lost water more quickly at Turin than at Col d'Olen.

The egg as a whole is continually losing water, while the embryo is continually gaining water, but at an ever-decreasing rate. The yolk receives water from the white during the first ten days of development. This transmission of water within the egg shows that the white has some of the functions of a water reservoir.

The water-content of the embryos of birds decreases with age. The same rule is followed by the embryos of man, guinea-pigs, rabbits, mice and cows. Measurements show that the process continues after birth in the case of rat, cat and dog. Several authors have concluded from these observations that growth is always accompanied by a decrease in water-content. Cramer found that the water-content of neoplasms (growths such as tumours) is 5 per cent higher than that of normal tissues, and concluded that a high water-content was connected with a high rate of growth.

Gray found the water-content of embryo trout remains constant at about 85 per cent. Bischov found that the water-content of a newly-born whole man is 66·4 per cent, whereas the content at thirty-three years was 58·5 per cent. The corresponding figures for the muscle only were 81·5 per cent and 75·67 per cent. This tendency of the water-content to decrease with age has been regarded by Ruzicka as the foundation of the ageing process, which he conceived as a continual tending of protoplasm towards stability, solidity, insolubility and dryness. Needham has suggested that the high water-content of young organisms may have quite another explanation. He cites the experiments of Skelton, who showed that bleeding in mammals produced a flow of water from the tissues into the blood. He found that the muscles contributed 10 per cent of this water, the liver 8 per cent, the intestines 2 per cent, the spleen 0·3 per cent and the connective tissues 78 per cent. Hence the decrease in water-content of mammal embryos with age may be due only to the decrease in the amounts of connective tissue. The decrease in water-content contrasts the embryos of mammals with those of fish. The trout embryo's water-content remains constant during a large part of the development and must be able to obtain water from outside itself, because the total amount of water in it is proportional to its weight, and its weight increases with growth. Gray showed that the newly fertilized trout egg contained enough solid for the construction of the finished embryo but not half enough water. He described the water-solid relations by the equation:

$$\text{Wet yolk} + \text{External water} = \text{Wet fish} + \text{Dry yolk used for}$$
$$(\text{1·0 gm.}) \quad (\text{0·7 gm.}) \quad (\text{1·56 gm.}) \quad \text{other purposes}$$
$$(\text{0·14 gm.})$$

Gray showed that the dry-yolk item was consumed in respiration to provide energy for the growing process.

Ranzi obtained a similar equation for the squid:

$$\text{Wet yolk} + \text{External} + \text{External} = \text{Wet squid} + \text{Dry yolk used for}$$
$$\text{water} \quad \text{ash} \quad \text{other purposes}$$
$$(\text{1·0 gm.}) \quad (\text{0·78 gm.}) \quad (\text{0·033 gm.}) \quad (\text{1·727 gm.}) \quad (\text{0·089 gm.})$$

Certain catfish offer interesting examples of embryonic water absorptions. The catfish of the genus *Bagrus* incubate their embryos in their mouths. Wyman noted in 1857 that the hatched embryos which had not yet left the parental mouth were considerably heavier than the undeveloped eggs.

Wolf found that phyllopod eggs could develop normally after fourteen years' drying in a desiccator, and Carpenter noted similar qualities in rotifer eggs. The embryos of these animals obtain nearly all their water from their environment. Their resistance to desiccation enables them to live in the periodically drying pools in which they are found.

The insufficiency of the original water-content of the eggs of water animals may explain some curious phenomena such as the rotation of the eggs of the pike. According to Kasansky, pike eggs rotate inside their containing membrane. The rotation may produce currents within the egg-case and help the absorption of water from outside.

Gray has commented on the evolutionary significance of embryonic water-absorption. The primitive vertebrate is generally regarded as the offspring of an aquatic animal. At some stage the first land animals must have begun to lay their eggs on land instead of in water. These animals would not have eggs similar to those of trout, because the trout embryo soon fills its enclosure and is necessarily born before it is highly developed and able efficiently to fend for itself. An exceptional development of the embryo before hatching requires a large space between the yolk and the enclosing membrane. Thus the reptiles and their descendants probably arose from an organism that laid eggs in water. The eggs did not hatch until the embryo had attained an advanced development. The space in the eggs of this organism which enabled the embryos of its aquatic ancestors to attain an exceptionally advanced development then changed in function. Whereas in water it had been a provision of space for expansion, on land it became a reservoir for water. The space around the yolk of the hen's egg is essentially a water-reservoir.

During the first nineteen days of incubation the chick embryo absorbs 10 cubic centimetres of water from the white of the egg. The white of an egg is thus the descendant of the primeval sea. Like the sea of blood in which the tissues of animals are bathed, the white of eggs is a preservation of the sort of environment in which the first eggs arose in evolutionary history. Living organisms have never departed from the principles of aquatic life. When they became adapted to life on land they accomplished the adaptation by taking the sea with them, within their eggs and within their own skins. All birds' eggs and the eggs of many reptiles contain enough water for development, but the eggs of some reptiles are partially deficient in water. These eggs deficient in water swell after being laid. The incomplete independence of the eggs of many land animals from external supplies of water illustrates an incomplete adaptation to life on the land.

Karashina has studied the various contents of the developing eggs of the Japanese marine turtle. It lays eggs of about the same size as ping-pong balls in damp sand above high-water mark. His results showed that the yolk absorbed water from the white, and that the eggs absorbed from the exterior a quantity of water equal to 42 per cent of their original water-content. Needham comments that these turtle embryos are left to find about one-third of the water they require, and this may explain the relatively small size of the eggs of many members of the turtle group. Cunningham has reported how a turtle from North Carolina in America has a special organ for wetting the earth in which the eggs are to incubate. These turtles build their nests in high ground sometimes at a considerable distance from the water. Usually the ground chosen is hard and dry, but sometimes a sandy beach is chosen. The turtle wets the ground with liquid from a special bladder, as the earth in

the hole is wet, while the surrounding earth is dry.
The function of this bladder had not previously been
understood. Cunningham analysed the liquid and
found it to be a dilute urine. During development
the eggs swell slightly. Needham remarks that the
behaviour of this turtle provides a notable link in the
evolutionary chain, as the turtle makes a special
effort to provide its land eggs with sufficient water,
but has not yet succeeded in making the provision
inside the egg.

The necessity of additional water during the development of eggs probably explains many examples of
curious rearing habits. Wheeler states that ants
'salivate' over the eggs in their communities and
suggests that the liquid may be absorbed. Weyrauch
reports that earwigs lick their eggs after laying them,
and that the eggs will not develop unless this is done.
There is a considerable amount of evidence to show
that many insect eggs will not develop properly without a humid environment. Harakawa found that at
a relative humidity of 10 per cent only 49.4 per cent
of the eggs of the Oriental peach-moth hatched,
whereas at a humidity of 65 per cent, 100 per cent
of the eggs hatched. It is known that Trinidad
froghopper eggs will not hatch unless the humidity
is over 90 per cent. Bodine has shown that the
water-content of the whole grasshopper egg rises
during development. Peacock made interesting experiments with the eggs of saw-flies which prove that
these eggs absorb water from the leaves of plants.
The saw-fly makes pockets in gooseberry leaves for
the accommodation of its eggs. If the stalk of the
gooseberry twig is immersed in water containing the
dye eosin, the dye passes into the leaves and thence
into the eggs, evidently being carried there by absorbed
water. When the eggs are removed from the pocket

at the beginning of development they develop normally by absorbing water from the air. Needham comments that the insects, unlike the female animals that themselves provide water for their eggs, seem to have solved the problem of living on the land by producing a sort of deliquescent egg which absorbs moisture from the atmosphere. The insects have an adaptation problem simpler than that of the larger animals because the air in restricted spaces is more moist, due to the effects of water absorption and capillarity. The crevices in which they lay their tiny eggs are naturally moist, and have a humidity much greater than that of the surrounding air. Hence the micro-climate of the crevice, as Needham expresses it, is much more humid than the main climate in which the insect lives.

Gray looks for the evolutionary ancestor of the higher animals among fishes whose eggs possess a watery membrane. The eggs of certain fishes have watery membranes whose function is protective. The eggs of all amphibians have membranes of this sort, apparently to protect the egg against predatory enemies. In some amphibians the membrane is sticky and does not obtain its full supply of water until the egg has been laid in water. Some amphibians that lay their eggs on land surround them by frothy material which becomes liquid and provides a liquid medium in which the eggs can develop. The female of one species makes a subterranean burrow and deposits a sticky substance. It treads this substance into a froth with its feet, and then lays its eggs in the froth. As the eggs develop the froth liquefies.

In this and other examples the water for the full development of the egg comes from the mother. The white of an egg is apparently descended from the watery protective layer on certain fish eggs. The investigator may search among the lower organisms for the first appearances of structures which in higher organisms have become of essential importance. The simple protective layer on the fish-egg may evolve into the very important egg-white which allows the independence of water in development necessary for colonizing the land. Hence the investigator may proceed from a consideration of the highly-evolved hen's egg to a search for insignificant layers on the eggs of fishes. He may follow the reverse path, and search for primitive characteristics persisting in the advanced product of evolution. If the egg-white is descended from a watery protective layer it might still possess noticeable protective qualities. Needham mentions that amphibian egg-jelly is remarkably resistant to the attacks of bacteria, and egg-white also is resistant. The protective function is perhaps persisting in the egg-white as a resistance to infectious disease.

The eggs of those strange animals, the duck-billed platypus or water-mole, and the spiny ant-eater, have curious properties that help to clarify the general evolution of eggs. The water-mole has affinities with the reptiles and the mammals; it is midway in the evolutionary order. It lays eggs, but when the eggs are hatched it suckles its young. It has no developed udders and lies on its back so that milk may ooze through the skin into a depression. The young suck from this pool. The water-mole, unlike a perfect mammal, is unable to keep its internal temperature very constant. Gray remarks that the eggs of this sort of animal have no true albumen layer or white. When the egg, which is mainly yolk, leaves the ovary, it is about 2 millimetres in diameter, but as it passes down the oviduct or tube from the ovary it grows enormously in size. The yolk swells up to a diameter of about 14 millimetres, so that the volume is increased 300 times. The egg evidently absorbs much water from the sides of the maternal tube. Thus the egg of the water-mole in contrast with that of the reptile absorbs water directly into its yolk from the walls of the maternal tube, while the reptile deposits the watery supply around the eggs after they are laid. In both cases the extra supply of liquid comes from the mother, but the water-mole has a more direct and highly evolved method of communication. In the higher mammals this method has evolved one stage farther, as the mother passes water from her blood not into the egg but into the embryo within her. Gray concludes that the land animals possessing backbones are probably descended from a fish-like ancestor, the walls of whose egg-laying tube were capable of producing a watery secretion. The secretion was at first used as a protective covering for the eggs and later became one of the essential characteristics that enabled the eggs of descendants to develop on land, by assuring them an adequate supply of water.

This survey of the ash-content and water-content of the eggs of the various orders of animals shows that comparative chemical study provides new perspectives of the course of evolution. Mere similarity of shape, such as that of the watery layer around the yolks of certain marine eggs and with that of the white of a hen's egg, is not usually sufficient evidence for an evolutionary connection. There must also be a persistence of function, or a traceable history of the change of function. The study of the chemistry of the egg directs attention to the functional aspect of organs and parts, what they do, and how they do it. From one aspect chemistry is the science of fuels, as the reactions of one substance with another are accompanied by the production of heat or cold. Chemical study naturally directs the student towards an aspect of living organisms which particularly exhibits their analogy with heat engines. The reactions of the substances in the organism correspond to the burning of the fuel within the engines. Hence chemical examination reveals a filled, inside view of the organism.

The fat-content of the various orders of eggs show interesting peculiarities. In general the ratio of fat to protein is higher in the eggs of land animals than in the eggs of water animals. For instance, the weight of fat in the hen's egg is about equal to the weight of protein. The eggs of the grass-snake have four times as much fat as protein, while the tortoise has about half as much. In cod eggs there is only one-twelfth as much fat as protein; herring eggs have one-eighth, and frog eggs less than one-third. The eggs of the dogfish and the sturgeon have relatively high fat-contents, as the fat equals the protein in the former, while in the latter there is one-half as much fat as protein. The high fat-content of the eggs of the sturgeon helps to explain the superiority of Russian caviar, which is made chiefly from sturgeon eggs.

A high fat-content in the egg like a high water-content, may be one of the characteristics that enables a water-animal to colonize the land. In order to become adapted to life on land a water-organism must not only arrange to obtain at all stages of development adequate supplies of the liquid to which it was previously accustomed; it must also become adapted to an immensely more rigorous climate. As Needham writes, it could no longer deposit its eggs in a cosmic thermostat, where the embryos could develop under uniform and constant conditions of temperature, pressure, gas-content, saltiness, etc. The sea presents an extraordinarily stable environment, in notable contrast even with fresh-water. Indeed, adaptation to a fresh-water environment may have been an intermediate stage towards adaptation to the land. Fresh-water is very much less constant than sea-water in temperature and composition. There is much less of it, and it is spread thinly over the earth's surface as rain. It drains through surface layers of varying composition and dissolves different salts in different places. Its slight bulk and proximity to the earth's surface exposes it to large temperature changes. The water in the seas is too immense in bulk for its composition to be noticeably changed even in millions of years, as the discharges of salt from the river-waters are too small to produce an effect in shorter periods of time. The severity of the fresh-water environment is illustrated by the before-mentioned small number of marine animals that have become adapted to it. The animals that left the sea had to have eggs capable of withstanding the rigours of the new climate, and indeed their departure from the sea depended on having eggs which already possessed protective characteristics capable of being developed and adapted to land conditions. The highly self-contained egg with a hard shell or tough case, which would prevent evaporation of necessary internal water supplies, with a white containing much water and yet resistant to disease, and a large fat-content, was a good solution, perhaps the only possible first solution, of the problem of embryonic life on the land. But the highly self-contained and protected egg has difficult internal problems. It must produce the embryo with little help or material from outside. It must be able to draw upon considerable sources of internal energy in order to accomplish a lengthy process of complicated construction. As fat is one of the most efficient heat-producing fuels, a good supply in the eggs of the more highly developed animals is to be expected. Needham has compiled a table of the number of eggs laid yearly by the females of various fishes. It is interesting to compare the fat-content of their eggs with the degree of self and parental protection with which they are endowed.

(After figures quoted by Needham.)

	Fat expressed as fraction of protein-content in egg.	Eggs per female per annum.
Cod (eggs simply discharged in sea) .	$\frac{1}{12}$	9,000,000
Carp (eggs scattered promiscuously over the submerged vegetation) . . .	$\frac{1}{9}$	400,000
Salmon and sturgeon (large eggs laid in holes scooped in gravel in shallow water)	$\frac{2}{5}$	2,000
Dogfish (eggs laid in hard cases anchored to sea-weed)	$\frac{1}{1}$	10

As the fish-eggs tend to be more protected the number laid increases and the fat-content tends to increase. The combination of high protection and high fat-content provides the conditions for a lengthier and more complicated embryonic development. When these are not provided the organism emits vast numbers of eggs into the sea, so that some may survive according to the laws of probability.

The eggs emitted by a female fish are fertilized by sperm emitted by a male. The sperm and the eggs combine together in the surrounding water. When organisms left water to live on the land they had to be provided with special mechanisms to enable their eggs and sperm to combine. Thus the mechanism of copulation was necessary to meet the conditions of life on the land. In ancient mythology Aphrodite Pandemos, the goddess associated with copulation, was represented as rising from the sea. Needham has remarked that she could with more exactitude be represented as rising from the land.

The copulatory mechanisms have been developed to suit the conditions of life on the land, and similar changes have occurred in the excretory mechanisms. The contents of a highly self-contained egg, such as the hen's egg, develop within an almost sealed enclosure. The complicated series of chemical changes must be accomplished within an enclosure impermeable at least to solids and liquids. Gases may pass through the egg skin and shell, and a certain amount of water in the form of vapour or gas, but no liquids and solids as such. The neatness of the chemistry of the embryo within a largely impermeable enclosure such as that of a hen's egg is astonishing. Compare the operations within the confined space of an egg-shell with the operations in a chemical laboratory. The chemist has arrays of bottles and taps. Liquids flow out of them often in unknown quantities into a small test-tube or reaction chamber. The test-tube is the centre where substances supplied from a multitude of divergent places meet. The chemist may draw upon unlimited quantities of reagents, energy, etc. The materials within the egg accomplish a large part of their reactions among themselves, and the reactions are an extraordinary example of finished chemical organization.

The chief materials in the egg are protein, salt or ash, carbohydrate, fat and water. The construction of the embryo out of these materials is accomplished with very little waste, but not entirely without. A perfect utilization of every original chemical atom in the egg is not expected, after considering the extreme wastefulness of human chemical laboratory methods. The absolutely perfect egg would resemble those mechanical or geometrical puzzles in which one figure is constructed from the divided parts of another figure. The rearrangement of the atoms of the original material in an egg to form an embryo is accomplished with an astonishing but not absolute completeness. It is one of the most impressive of biological phenomena. Even the incomplete aspects have peculiar interest. The chemical rearrangement of the atoms in the fat, carbohydrate, ash and water can with the assistance of some oxygen absorbed through the shell be accomplished with a very high degree of completeness, but the rearrangement of the atoms in the protein material, which contains awkward, lethargic nitrogen atoms, is not so complete. The protein in eggs cannot be converted into embryo without the production of some waste substances containing un-usable nitrogen, which is accumulated as a constituent of ammonia, urea and uric acid. Ammonia is a gas compounded of hydrogen and nitrogen only, and very soluble in water. Urea contains hydrogen, nitrogen, carbon and oxygen, and is also soluble in water. Uric acid is a more complicated combination of the same substances that is slightly soluble in water. The researches of Needham and others have shown that in the hen's egg the production of ammonia, urea and uric acid is not in steady proportion. When the ammonia production is expressed as a percentage of the dry weight of the embryo, it is found to decrease steadily from at least the fourth day of incubation until it has almost disappeared at the time of hatching. The production of urea is not noticeable until the fourth day and reaches a maximum at the eighth or ninth day. The production of uric acid is not noticeable until the eighth day and reaches a maximum at the eleventh day. These figures refer to the variations in the ammonia, urea and uric acid expressed as percentages of the dry weight of the embryo. As the embryo is continually increasing in weight, and urea, when it is produced, is produced more rapidly than ammonia, and uric acid, when it is produced, is produced more rapidly than urea, there are very marked differences between the total productions of each of these substances during the period of incubation. In fact, 91·5 per cent of the nitrogen excreted by the embryo during its development is in uric acid, 7·58 per cent is in urea, and 1·07 per cent is in ammonia. Yet at the fourth day nitrogen excretion is restricted to ammonia, while at the fifth day excretion is restricted to ammonia and urea, and the uric acid production is still very small on the seventh day. Needham has summarized these figures in a table.

(After Needham)

	Percentage of the total nitrogen excreted during development of embryo.	Time of maximum production. Days.	Percentage of nitrogen in compound.	Molecular weight of compound.
Ammonia . .	1·07	4	82·3	17
Urea . . .	7·58	9	46·6	60
Uric acid . .	91·35	11	33·3	168

In the first stage of growth the embryo excretes ammonia which has a high content of nitrogen and a simple structure. In the second stage it changes its excretory end-product into a more complicated substance, and in the last stage it excretes chiefly uric acid, which is still more complicated, and is the form of nitrogen excretion chiefly used in the adult bird. Needham has given an interesting comparative review of this information concerning the nature of the excretory products of various organisms. He remarks that students beginning a course of biochemistry always ask why some animals such as man and mammals excrete chiefly urea, some ammonia and some uric acid. He writes that they have not received a reasonable reply, though he believes one can be given.

Urea could not be the chief excretory product in the hen's egg because it is very soluble and diffusible and hence unsuited to the organization of life in an impermeable enclosure. Uric acid is more suitable because it is difficultly soluble and stays where it is put. Life in a limited space is much easier if the end-products are not very soluble and reactive. Once it or its insoluble salts are deposited in a place they are not too easily removed, as sufferers from gout know. The uric acid or urate crystals are deposited in the joints and are not easily dissolved away. The chick embryo has to pay for the advantage of using uric acid by providing a large water transport system for it. Fiske and Boyden have calculated that 15 per cent of all the water in the egg is employed in the transference of the five-thousandths of a gram of uric acid present on the eleventh day from the embryo to a corner where it is harmless. So much water is needed to dissolve so little uric acid.

If urea were used by the chick embryo as its chief excretory end-product its diffusibility would cause the embryonic tissues to be soaked in it; the concentration being about 165 milligrams in each hundred grams of embryo. The normal urea-content for human blood is 25 milligrams per hundred grams of blood. In severe renal obstruction or nephritis it rises above 100 mgm., and may reach 300, but 165 is already of pathological magnitude. Hence the chick embryo that used a urea excretory system would, as Needham remarks, be suffering from a constant headache and other symptoms before hatching, and natural selection would hardly have preserved it for our entertainment. It has avoided these consequences by adopting a uric acid excretory system. It has already been mentioned that the proportion of protein to fat is much higher in the eggs of water than in land animals. The embryos of water animals derive the energy necessary for the chemistry of their growth from protein rather than from fat. For instance, the chick embryo obtains about 90 per cent of its energy from the oxidation of fat, while the frog embryo obtains about 70 per cent of its energy from protein. The eggs of the silkworm, which is a land animal, obtain about 60 per cent of their energy from fat, whereas the eggs of plaice get about 90 per cent from protein. Perhaps the change from water to land conditions forces the embryo to change from the use of protein to fat as fuel because the products of protein combustion are so much more difficult to dispose. The difficulty of disposal within the impermeable egg has forced the adoption of the uric acid excretory system.

Animals that have progressed beyond the first colonizers of the land, who accomplished colonization with the help of the laying of suitable eggs, and have evolved wombs in which embryos may remain in intimate fluid communication with their mothers, use a urea-system of excretion. The embryo of the mammal can discharge its waste nitrogenous products into its mother's blood system. Hence the mother's blood system provides it with a liquid environment, as the primeval sea provided its fishy ancestor with a liquid environment. The mammal embryo can, in virtue of its liquid environment, make use of the soluble urea. Here is a possible explanation of why fishes discharge the waste nitrogen from their protein in urea while birds and reptiles discharge it in uric acid, and mammals have returned to urea. Colonization of the land was achieved through the impermeable egg and this necessitated the uric acid excretory system. The sort of animals that existed in early geological times may not have been able to emerge from the sea without the invention of a uric acid excretory system, though this does not appear to be necessary. In order to colonize the land the early amphibia had to evolve the impermeable egg, or viviparous mechanisms of bearing grown offspring. The second line of evolution is conceivable, but they apparently chose the first as the route to the mammals.

The use of the uric acid system has handicaps. Ammonia is in itself a very efficient end-product because it contains a high percentage of nitrogen and takes no carbon away with it. It also is formed without the absorption of much energy. Urea is less efficient in itself because it cannot take nitrogen away with it without taking half as much carbon also, and its formation requires almost twice as much energy as the formation of ammonia. Uric acid in itself is still less efficient because it takes as much carbon as nitrogen with it, and its formation requires five times as much energy as the formation of ammonia.

The decision whether an organism uses urea or uric acid appears to depend on the circumstances of embryonic life. The adult human being excretes chiefly urea because the human embryo can use urea, which is more efficient than uric acid. The bird excretes uric acid because its embryo cannot use the more efficient urea.

Which organisms use the very efficient ammonia, and why is it restricted to them?

Needham has collected the data concerning the constitution of animal urines. Some representative figures are given in the table, extracted from his collection.

Sort of animal.	Egg developed in water or land environment.	Substance through which nitrogen excretion is done, and percentage of the total urinary nitrogen.		
		Ammonia.	Urea.	Uric Acid.
Protozoa: Paramoecium .	W	—	90·0	—
Didinium . .	W	90·0	—	—
Worms: Sea-mouse . . .	W	80·0	0·2	0·8
Earth worm . .	L (?)	20·4	38·1	trace
Insecta: Silkworm . . .	L	—	—	85·8
Gastropoda: Sea hare . .	W	33·5	8·7	4·6
Slug . . .	L	4·6	70·8	6·9
Fish: Dogfish	W	7·3	80·7	0·2
Carp	W	56·0	5·7	0·2
Sole	W	53·0	16·6	—
Lung-fish	W	41·2	18·5	0·8
Amphibia: Frog . . .	W	15·0	82·0	trace
Reptilia: Turtle . . .	W	15·3	39·0	18·8
Alligator . . .	W	75·3	7·2	13·1
Tortoise . . .	W	—	90·0	trace
Snake	L	—	—	80·0
Lizard	L	—	—	91·0
Birds: Hen	L	1·5	0·9	70·0
Duck	L	3·2	4·2	71·9
Mammalia: Man . . .	W	4·3	87·5	0·8
Whale . .	W	1·5	90·0	3·0
Dromedary .	W	12·3	55·5	0·3
Dog . . .	W	3·0	89·0	1·0
Monotreme: Spiny Ant-eater	(?)	6·9	81·2	—

An examination of these figures shows that the elementary marine animals preferably use ammonia, while the higher marine animals use a urea excretion. The water reptiles preferably use urea, while the land reptiles use uric acid. The birds also use uric acid, and the higher mammals urea.

The very elementary animals first used ammonia, and then preferred urea. Why did they change to a substance in itself a less efficient carrier? Because ammonia is strongly alkaline. Unless it is removed rapidly acid must be produced to neutralize it. Ammonia can be removed rapidly without the production of auxiliary acid only when the simplicity of the structure of the organism allows the ammonia to diffuse directly after formation into an unlimited supply of adjoining water. The simplicity of structure necessary for the use of ammonia exists only in the lowest animals. In more complicated animals auxiliary acids, useless except as an assistant in ammonia excretion, would have to be manufactured. The efficiency of ammonia as a carrier of nitrogen would be discounted by the waste entailed in the production of acid necessary for the removal of excessive alkalinity due to the ammonia. The majority of the water animals soon turned to urea as the nitrogen excretion transporter. When some of these began to colonize the land they changed once more to uric acid because this allowed the use of impermeable eggs for the protection of developing embryos from the rigorous land climate and the storage of food supplies. The change from urea to uric acid occurred in the reptiles, and also in the insects which preceded the reptiles in the invasion of the land. The insects and the reptiles were presented with similar embryological problems when they began to invade the land, and they solved some of these problems in the same way. This is an example of the biological phenomenon of convergence, which is sometimes seen when very different sorts of organisms follow parallel lines of evolution because the conditions of evolution on each of the lines happen to be the same or similar.

The use of ammonia, urea, uric acid, and urea again, as the medium of the disposal of the waste nitrogen products of growth corresponds to stages in evolution. Ammonia excretion is the most primitive, then urea, then uric acid, which must follow unless the evolution of a system of embryonic growth within a parent animal has been successful, allowing the embryo to discharge soluble waste products into the maternal blood-stream.

The researches on the production of ammonia, urea and uric acid in the hen's egg show that in the early stages the chick embryo uses ammonia excretion, in the middle stage urea excretion, and in the later stages uric acid excretion. This is one of the most remarkable examples of the biological phenomenon of recapitulation. The development of mammalian embryos through stages in which their shapes resemble those of primitive organisms, fishes, lower mammals, etc., is well known. This recapitulation of the mechanisms of living matter used at the various stages of evolution occurs in the development of the chemical mechanisms of an embryo; recapitulatory phenomena are evident in the chemical besides the morphological aspect of living matter.

The variety of chemical facts provided by the comparative studies of the chemistry of embryonic excretion had no clear significance until Needham propounded his brilliant interpretation, and the existence of uric acid excretory systems was entirely without explanation. His explanation is that the evolution of many animals living on the land would have been impossible without it.

III

The recapitulation or repetition of geometrical shapes and chemical mechanisms by the developing embryo is at the first sight one of the most striking phenomena in biology. The most complicated mammals appear to be the product of a unique line of evolution, and each animal seems to be the culmination of the evolution of its predecessors. If embryonic growth were strictly recapitulative there would be one line of evolution only. Corresponding to any specified point on the line, there would be only one possible organism. A point near the beginning of this imaginary line could correspond with, say, amoeba, and no other sort of animal. A point farther along would necessarily correspond with, say, the fish, and much farther along with, say, the dog. At this latter point one sort of animal only would be found, not a variety of animals of entirely different structure with exactly the same abilities. A perfect recapitulation would imply that evolutionary development is restricted to one line of advance. It would enormously enhance the uniqueness of the living process. The conception that there are many ways of constructing an organism with the abilities of a dog is not unreasonable. Nor is it unreasonable to suppose that the material of an egg could develop into a chick by a multitude of alternative routes. In fact, the germ of the egg always develops through a series of shapes, including one with strong resemblances to a fish. Why should it proceed through a standard series of shapes and chemical mechanisms? Recapitulation is a striking phenomenon because it seems to imply that living matter is confined to one main line of evolution, and that a developing organism must always pass through the use of the shapes and mechanisms achieved by its predecessors. As this view implies the line of evolutionary advance is unique the observer begins to associate a unique entity with the unique line. Evolution is pictured as the journey of an abstract entity named life along a unique road through the material world. Whenever this independent abstract entity retraces the road it has previously made through the material world, the road from the amoeba to man, it remembers what it was like as it passes each point of its previous journey, and immediately organizes itself into a copy of the figure and processes it then had. The perfect recapitulation theory makes the science of embryology a gallery of ancestral portraits, and when it was fashionable many embryologists examined the stages of embryological development in order to discover portraits of man's ancestors.

A balanced review of the facts of embryology shows that the early enthusiasms for the recapitulation theory were not well-founded. When observers refrained from looking for resemblances and concentrated on complete objective description, embryos possessing resemblances often had more differences than resemblances. The recapitulation theory proved to be only partially true. When it is reduced to a description of rough repetitions it loses its striking attractiveness. The idea of that abstract entity named 'life' journeying along the road of evolution becomes vague, and the theory of recapitulation loses its usefulness to the animistic thinker who desires to conceive the living organism as a material configuration directed by an external spirit. Though the completer generalizations of the recapitulationists are incorrect, the assembly of repetitive facts in embryonic development is considerable and important. The problem of the production of the shapes and mechanisms of embryos at each stage of development is part of the chief problem of biology, which is the discovery of the mechanism of organization. How are multitudes of atoms organized into bodies possessing the properties of ten-day-old chick embryos, or dogs, or professors? Within this general problem of organization, why are there discontinuous elements in the development of embryos, as the repetition phenomena show? Very little is known concerning biological organization. Perhaps that is why the subject is particularly attractive to philosophical speculation. Backward sciences legitimately stimulate speculation because of the necessity for inventing methods of research which will remove the backwardness. They are also often attractive to speculators because the insufficiency of factual data makes erroneous views more difficult to disprove.

The study of the chemistry of growing organisms tends to make animistic theories of the organization of living matter less tenable, because the student concentrates his investigations on to the inner structure and functions. As he diverts his attention from the study of external shape, from morphology, he becomes emancipated from the types of speculation which an exclusive study of external shapes engenders. He loses the desire to populate the empty shapes. The study of the internal structure and mechanism engenders its own philosophy. The majority of the chemical embryologists at present making definite contributions of new knowledge believe that the nature of the organization of living matter is not essentially different from that of non-living matter. It is more complicated, and the complications bring into action effects not seen in simpler material structures. According to this view, the structure and mechanisms of living matter follow in the final analysis from the natural structure of atoms and molecules themselves. The structural organization of atoms depends on the fundamental properties of matter, space and time. The experimenter engaged in finding new facts concerning the chemistry of embryology and growth naturally tends to connect the properties of growth and organization with the agents with which he works. He suspects the shapes and mechanisms of living organisms are a product of the properties of the atoms of which they consist. He speculates that the nature of the shapes and mechanisms is based on the shapes and interactions of the atoms and molecules in the living matter.

The general researches into the problem of biological organization contain at least two lines of work which promise the discovery of a little firm ground under the clouds of speculation. These lines are the 'organizer' studies started by H. Spemann; and the X-ray studies of the structure of biological material that are an offshoot of the X-ray analysis of the structure of crystals. In England the study of the structure of biological material by X-ray photography has been brilliantly advanced by W. T. Astbury.

In 1924 H. Spemann and Hilde Mangold discovered that young embryos have regions which control the organization of adjacent growth. If a piece of tissue was cut out of this region and grafted into another embryo, it was able to organize the surrounding cells in its new environment into a more or less complete new embryo. For instance, the transplanted tissue might be grafted on to the belly of an embryo where it would organize its own and the neighbouring cells in the skin of the embryo's belly into a complete head.

The artificially-made monstrosity had a complete normal body, with an extra head growing on the middle of the belly. Otto Mangold, who is one of the leaders of the Spemann school, has made many experiments with the eggs and embryos of frogs. He transplanted a portion of the brain-cells of a two-day-old embryo on to the breast of another embryo. When the transplanted brain-cells are taken from an embryo of two days' age they are assimilated by the skin of the breast of the host embryo, becoming incorporated as normal breast tissue, and losing their original identity as brain-cells. If the transplanted brain-cells are taken from an embryo of four days' age they grow on the skin of their embryonic host, but they are not incorporated as cells of the skin. They preserve their original identity as brain-cells and grow into a brain. Hence the embryonic host is provided with an extra brain growing on its breast. Besides preserving their original identity as brain-cells and growing the four-day-old transplanted cells have a still more remarkable power. They can enable the cells in the skin of the host's breast to assist in producing entirely new cells corresponding to parts of the brain not transplanted. For instance, the embryonic cells from which the eye develops may have been transplanted, but other cells corresponding to other parts of the head may have been left behind. Nevertheless, the extra brain and head appearing on the embryo's breast has parts and features whose primitive cells were not transplanted. In these experiments the transplanted cells are taken from the brain of one embryo and planted on the breast of another, so that there is a difference in location on their source and their host. If corresponding groups of cells are exchanged between two embryos, so that one group of cells is planted into the hole left by the other,

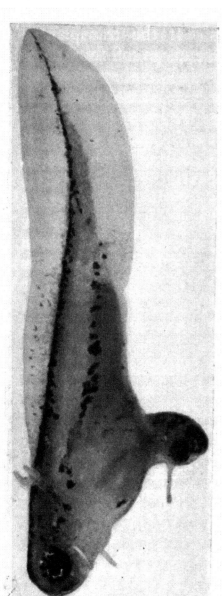

(a) A part of one developing newt egg was transplanted into another developing newt egg. The larva from the host-egg is shown thirteen days after the operation. The transplanted speck of tissue has induced another head containing a brain, nose, one cyclopic eye with lens, one hearing organ, sense-buds, green and black pigments, etc., and has incorporated cells from the host into the structures. (Transplant from gut of early neurula into blastocoel of early gastrula.) (O. Mangold.)

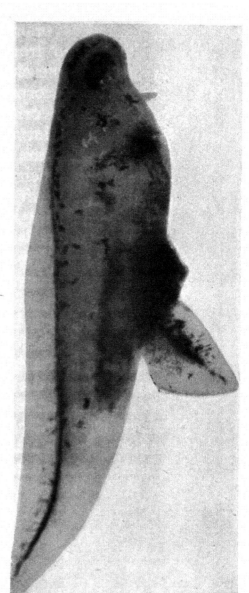

(b) One newt egg received an implant from another newt egg. The picture shows the larva into which the host-egg had developed during the eight days succeeding the operation. A second tail has been induced, including spine-marrow and pigment. (Transplant from medullary plate of early neurula into blastocoel of early gastrula.) (O. Mangold.)

both embryos can grow into normal adults. The power of induction or direction, by which transplanted cells may cause the adjoining cells in their new environment sometimes to fall in line with their pattern and take part in the formation of organs and limbs which would not have appeared without the arrival of the transplanted cells, resides at first in the outer part of the embryo within the egg. As the embryo develops the power goes into its internal tissues. The frog eggs used in these experiments have a small groove. The hump or edge of the groove appears to contain most of the cells involved in directing the organization of the growth of the embryo, for if it is removed the embryo grows up into a shapeless mass.

Interesting results may be obtained by grafting eggs together. When an egg is fertilized and begins to grow it soon divides into two cells, forming a dumb-bell shape. Each of the pair of cells divides into two cells, making a cross containing four cells. It is possible to graft together two embryos in the dumb-bell stage so as to make one embryo in the cross form containing four cells. The grafted embryo may develop into a giant normal, triplet or quadruplet embryo. If the pairs of cells grafted together are from different varieties of animal, the manufactured embryo may have a coloured leg from one variety and a peculiarly shaped leg from the other variety, and so on.

Spemann grafted tissues from embryos of different colours from different species. As the graft grew into a new embryonic organ on its host, or was incorporated by its host, the difference in colour between the original transplanted cells and the cells of the host enabled the division between the cells of the transplant and of the host to be seen accurately during the period of growth.

Brachet showed that the fragment of tissue containing the organizer could have effect only when it was in continuous contact with the surrounding cells. If a cut is made in the tissues near the organizer fragment the organizing power cannot jump across the cut. The cells beyond the cut are not organized into a head or spinal axis. Hence the organizer is an entity which requires the presence of cells in order to be transmitted. In this respect it diffuses through tissue like some sort of fluid.

It also takes time to travel. Spemann found that the cells near the organizer fragment came under its direction before the more distant cells.

The experiment in which an organizing fragment from an embryo of a certain species was grafted into an embryo of another colour and species shows that the organizer is not specific to a species. One might have expected that the agent which organized the cells of one species into its typical form would be different from the organizing agent in all other species. Geinitz investigated this point by trying grafting experiments with embryos from increasingly different species, and showed that grafts could be effective when they came from embryos of different biological genera, i.e., from embryos of animals of quite different types.

The organizing power possessed by these special regions in embryos need not be exerted directly by original tissues from these regions. If other tissue is grafted into one of these regions it becomes endowed with the organizing power.

The organizer may retain its effectiveness for relatively long periods. Mangold found that a piece of brain from a free-swimming larva, i.e., an animal that had developed beyond the embryonic stage, could still organize a structure out of the cells of an early

PLATE XI

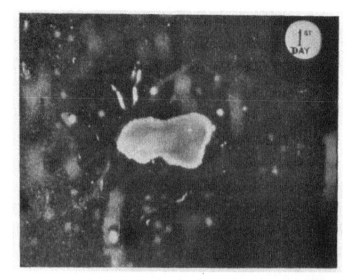

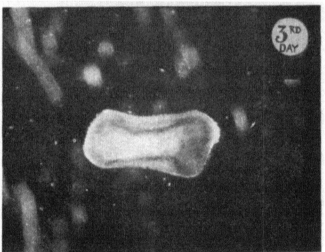

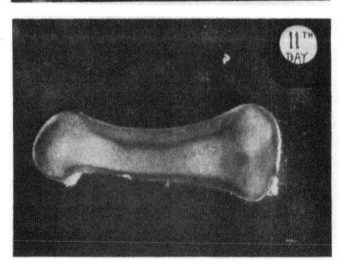

The thigh-bone of a fowl cultivated from a pre-thigh bone taken from an embryo five and a half days old. This bone grows to the appropriate shape in a suitable preparation, in spite of being detached from the body to which it belonged. The three pictures are from Dr. R. G. Canti's film made of the bone cultivated by Dr. Honor Fell. (R. G. Canti and H. Fell.)

embryo on to which it had been grafted. As the lines of organization of the cells of the larva had been determined at a much earlier date in its life, the presence of the organizer was presumably not necessary yet it still remained.

The organizer tissue may be subjected to the most violent treatment without losing its organizing power. It can be narcotized, crushed, dried, frozen or boiled, and yet retain the agent which enables it to direct the organization of neighbouring cells. Holtfreter has shown that parts of the newt embryo, which do not normally possess the power, gain it after being boiled.

When a group of cells has been set off by an organizer to develop into an organ they will continue to develop into the organ even if they are removed from the rest of the embryo, provided they are kept alive in a suitable nutritive medium. Fell and Robison have taken from the embryo of a chick the group of cells which ultimately develop into the thigh-bone. These were placed in a glass vessel containing a nutritive liquid. They gradually grew from an almost shapeless mass of pre-cartilaginous material into true cartilage with a little bone of the thigh-bone shape. Stangeways and Fell showed that the limb buds of chick embryos have little power of developing their various parts when isolated in nutritive media, but that the eye-buds are able to develop all the constituents of the normal eye when isolated, though they do not increase much in size.

Waddington, Needham and Needham have prepared purified extracts from the crushed tissues of newt embryos, which still retain the organizing power. The structure of the cells of the tissue was entirely destroyed by the crushing, and granules and other débris were removed from the crush by centrifuging.

(A centrifuge is a drum which can rotate very rapidly. The heavy particles in a liquid in the drum are hurled towards the outside, owing to centrifugal force, so that suspended particles are cleared from the body of the liquid and accumulated on one side.) The clear liquid obtained from the crush by the centrifuging may be coagulated by pouring on a warm plate. This solid material will organize cells when implanted into an embryo of suitable age.

If the newt embryo tissue is ground with anhydrous sodium sulphate and the product is treated with petroleum ether the solution contains an effective organizer. Petrol-ether extracts of the viscera of adult newts also gave some evidence of activity, which is in accord with the experiment in which Mangold showed that larvae or post-embryonic animals still contained the organizer. Waddington, Needham and Needham have tried to obtain similar organizing phenomena in embryos by implanting in them substances such as agar and egg albumen, but these gave no effect. This shows that the results obtained from implantations with the purified extracts from embryos are not due to mechanical stimulation, to cutting and pricking which occurs in the implanting operations. Consequently, they conclude that the organizing agent in the embryos of amphibians such as newts and frogs is a definite chemical substance, soluble in ether and probably of a fat-like nature.

Thus the organizer which directs the cells of embryos to arrange themselves in the pattern of their parents appears to be a definite chemical substance, and probably the same substance in all animals. Without it the cells of the embryo grow up into a shapeless mass. The organizer does not seem to force the cells to grow into the pattern of a head or a spinal axis or an eye; the cells themselves seem to possess their own power of growth. The organizer merely sets off in a trigger-like way their growth into the appropriate pattern.

In the future this potent chemical substance will very probably be prepared in the laboratory and placed in a bottle beside the various vitamins and hormones, which, to the astonishment even of scientists, are now available in labelled bottles on the shelves of chemical laboratories.

The organizer is present in the embryo from the beginning, but at first in a masked state, since it can be set free by boiling the uncleaved egg. Normally, it is set free between the second and fourth days in newt embryos. Thus the supervision of the structure of the body depends at one stage directly on its chemistry. If the correct chemical processes do not go on the embryo remains shapeless.

How should a chemical substance possess such a remarkable organizing power? Perhaps the structure of the molecule of the organizer allows it to act as a frame or scaffolding upon which the cells are assembled. A magnet orientates a haphazard pile of iron filings: perhaps in some analogous way the chemical organizer orientates a more or less haphazard pile of embryonic cells. W. B. Hardy has discussed the property possessed by some molecules of orientating themselves in chains so that they can have an effect at a distance. The researches of Astbury on the structure of fibres have raised the problem of the relations between the form of organisms and the form of the chemical molecules out of which they are made, so these will now be described.

IV

Rutherford has said that Röntgen's discovery of the X-rays in 1895 was the most stimulating discovery of recent physical science. It gave a profound psychological shock to physicists as it revealed the existence of a physical agent of entirely new properties. The nineteenth-century physicists were becoming encysted within stale physical conceptions and circumscribed in the use of their investigative imagination. The X-rays revealed new agencies and new regions of phenomena. With their help the existence of the electron was soon demonstrated and radioactive substances were discovered in the attempt to find other rays of similar properties. The most striking property of the X-rays was their penetrating power, which led immediately to remarkable applications in medical science. In spite of all the activity stimulated by the discovery of the extraordinary rays their nature remained unsettled for seventeen years. They were not finally proved to be wave-radiations similar to light, though of much shorter wave-length, until 1912, when von Laue showed they could be made to give effects characteristic of the interference of waves. When waves pass through or are reflected from parallel slits the various emerging wave-trains interfere. The effect may easily be demonstrated for waves of light by allowing a beam of rays to fall on a sheet containing two fine slits or pin-holes near together. If the screen is held at a suitable distance behind the sheet so that it receives the light coming from the illuminated pin-holes, it will bear the image consisting of alternate light and dark bands. The points on the screen directly opposite the illuminated pin-holes may even be dark. How this may occur can be seen from Fig. 27.

A and B represent the holes in the sheet, and C the point on the screen immediately opposite A. Some light will come to C from both A and B. If the distance AC may be such that the waves in the light from A have to undulate five and a quarter times before they reach C, whereas the waves from B may have to undulate five and three-quarter times because BC is longer than AC. Hence the waves from B will be passing through their trough when they reach C, while the waves from A will be passing through their crest. Hence a particle at C under the influence of the two waves will be torn equally in opposite directions. It will not be able to vibrate, so C will appear dark. On the other hand, the point D will appear light, as the distances AD and BD are equal, and therefore the number of the undulations along AD and BD are the same, so that the waves

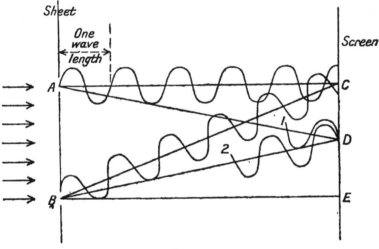

FIG. 27.

arrive at D both moving in the same direction. The point D appears four times as bright as if illuminated by pinhole A or pinhole B alone. Whether C is bright or dark depends on the numerical relations between the distances AC and BC, and the wavelength of the light. In order to obtain darkness at C there must be a difference of half a wave-length between the lengths AC and BC, in fact the difference is $5\frac{3}{4} - 5\frac{1}{4}$ wave-lengths. If the distance AC between the sheet and the screen is fixed, and the wave-length is small, the alternate bright and dark spots will be close together and hence indistinguishable unless the distance AB between the holes is extremely small. The early experimenters could not think of any arrangement of pairs of holes close enough to give bright and dark spot interferences when X-rays passed through. Arrangements which gave fine interference effects with light gave no result with X-rays. Then von Laue thought that natural crystals, which consist of atoms regularly arranged in rows, might provide such holes. The holes between the atoms might act for X-rays as the pin-holes act for light. The experiment was tried and succeeded, and the wave-nature of X-rays was proved. The seventeen-year discussion of the nature of X-rays was settled. This famous experiment was devised to establish the nature of X-rays; it was inspired by purely theoretical interests. But it was clear that if crystals could help to establish the nature of X-rays, X-rays might be used to investigate the properties of crystals. W. H. Bragg and W. L. Bragg developed this application, and H. Mark used the technique for examining the crystal-structure of materials used in industry, including plant and animal products. O. L. Sponsler and W. T. Astbury have been notably active in extending the technique to materials produced by living organisms.

Before the X-ray methods of investigating the structure of matter had been developed the crystalline state was believed to be rather unusual. The majority of substances were supposed to be amorphous, that is, without structure. This supposition was based on microscopical examination. Many substances when examined under the most powerful microscope showed no marks of ordered structure. For instance, substances such as skin and hair were believed to be entirely non-crystalline and made of molecules bunched together without exactly repetitive patterns. X-ray investigation showed this belief was incorrect and that in fact entirely non-crystalline materials are very difficult if not impossible to find. Ordinary microscopes failed to reveal the crystalline structures in nearly all solid materials because the constituent crystals were too small to be seen by visible light. X-rays can detect minute regularities far beyond the range of microscopical visibility. For instance, a fibre of silk reveals no structure under the microscope. It appears transparent and without visible structure, but when a suitable beam of X-rays is directed on to it and the emergent beam is photographed the photograph shows rings of black spots due to interference. Evidently the silk-fibre has an invisible crystalline structure. Silk consists of a protein named fibroin, which is exuded from the spinning gland of the silkworm. The X-ray photograph shows silk is not even a random conglomeration of crystals of fibroin but an ordered structure of crystals. Moreover, it shows that the crystals are all lying along the same direction and that this direction is the same as that of the axis of the fibre. Similar information concerning the structure of ramie fibres is given by X-ray photographs. Ramie or China-grass is used in the manufacture of mats. It is a vegetable fibre constructed of cellulose crystals lying longitudinally. The cotton fibre is also constructed of tiny long thin crystals lying side by side. As the cotton fibre does not lie straight, but has a natural twist, the cotton crystals do not lie parallel to the axis of the fibre, but inclined at a constant small angle. This feature is also shown in the X-ray photograph. When the cotton fibre is stretched the spiral becomes extended and the cotton crystals become aligned more nearly parallel to the axis. This phenomenon due to straining is shown by the X-ray photograph.

Thus the X-ray technique will show whether a material is crystalline; the arrangement of the crystals; their size; and the arrangement of larger units in which they are assorted.

Wet fibres give much the same X-ray pictures as dry fibres. This shows that water does not disturb the shape of the fundamental units of the fibres. In wetting it is absorbed on and between the crystal units, but not within them.

Astbury writes that the crystals in the ramie fibre are about one five-millionth of an inch thick and fifteen times as long. One ounce of ramie contains about 10 million million million crystals with a total surface area of 163,000 square feet. Hence for the absorption of water the internal surface of a fibre is vastly more important than the external surface. This explains why fibres such as cotton and wool can absorb upwards of a third their weight of water, and why a wool-fibre takes about a quarter of an hour to become thoroughly wet, while days are required to remove all the water from it by drying agents such as phosphorus pentoxide. The molecules of water are secreted between the surfaces of the unit crystals in the interior of the fibre and cannot easily be drawn out.

The X-ray photographs show that the unit crystals of fibres are about twenty times as long as they are broad. When the fibre is wetted the crystals obtain a uniform layer of water all over. This has a curious effect. A small cube of fibre will expand sideways about twenty times as much as lengthways, because there are twenty times as many lateral as longitudinal surfaces. The long thin structure of the constituent crystals causes surfaces to occur twenty times as frequently in the lateral as in the longitudinal direction. This is confirmed by the fact that wool fibres, for example, may swell 18 per cent in diameter but not more than 1 per cent in length when thoroughly wetted. This phenomenon explains why tent-ropes have to be slackened in wet weather, though rope-fibres become longer when they are wet. A rope consists of fibres wound in a spiral. When the fibres are wet they expand, and the radius of the spiral expands. But the radius of a spiral cannot be increased without shortening its length, if the number of turns remains constant.

Hence the necessity for slackening tent-ropes before wet weather arises ultimately from the shape of the crystal units of the rope fibre, as shown by X-ray photographs.

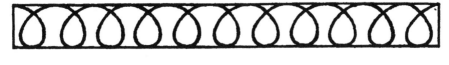

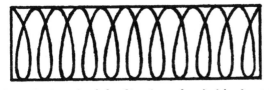

FIG. 28. The length of the fibre in each spiral is about the same, but the length of the thicker spiral is shorter.

The existence of long, thin units in the structure of plant fibres was suggested by the botanist Nägeli fifty years ago on evidence provided by the swelling properties of certain biological materials. He named the units 'micelles', a word derived from the same root as mica, the transparent material resistant to heat which easily splits into flakes.

Artificial silk is made from the long thin crystal units in wood-fibre. The artificial silk manufacturing process consists of taking lumps of wood to pieces, into their ultimate crystalline units, and making them into a thick solution. The units are arranged in longitudinal order again, as in a natural fibre, by squirting the solution through a pin-hole. X-ray photographs of artificial silk fibres show the characteristic long thin unit crystal structure.

The chief constituent of vegetable fibres such as wood, cotton, flax, ramie and hemp is cellulose. Chemists have shown that cellulose is probably produced by plants from a sugar named glucose. When a molecule of glucose is deprived of two atoms of hydrogen and one of oxygen it leaves a group of six atoms of carbon with ten of hydrogen and five of oxygen. Analyses of cellulose show that carbon, hydrogen and oxygen always appear in these proportions, but it is impossible to say whether a molecule of cellulose contains one or more of these groups. It appears to consist of an indefinite number of such groups strung together. Thus cellulose does not form a definite molecule because its fundamental unit groups are able to stick together in long chains. The vegetable fibres are fibrous because they are made of a chemical substance which is fibrous in its molecular structure. The size of the unit groups of carbon, hydrogen and oxygen atoms can be deduced from chemical measurements. It can also be deduced from the X-ray photographs of cellulose fibres, and the two results agree. The strings of atomic groups in cellulose molecules are usually about a hundred times as long as they are broad, and the crystals in vegetable fibres are merely neat bundles of these strings of grouped atoms.

The animal fibres, such as silk, wool, feather, nail and horn, are built of stringy protein molecules. Chemical research has shown that protein molecules are made of strings of groups of atoms. The length of these groups is always the same but their breadth varies. Hence a protein chain is not so regular as a cellulose chain. Its length is marked by joints at regular distances but the portions between the joints vary in breadth. The difference is illustrated by Fig. 29.

As there are many different protein group units of the same length but differing breadths, the protein chains can be made up in many different ways. In Fig. 29, four groups are arranged in the order 1,2,3,4, 1,2,3,4, ..., etc. They might be arranged in the order 1,3,4,2, 1,3,4,2, 1,3,4,2, ... With four different unit groups twenty-four sorts of chain are conceivable. With ten different unit groups 3,628,800

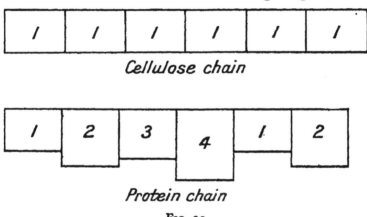

Cellulose chain

Protein chain

FIG. 29.

sorts of chain are conceivable, and yet the percentage chemical composition of all of these proteins would be the same. Their differences would be due entirely to different arrangement.

This is one of the most interesting ideas offered to biology by the students of organic chemistry and X-ray analysis.

The differences between species of animals may arise from changes in the order of the groups in protein chains. Suppose that the order of the groups in a protein chain in the dividing chromosomes of an egg that has just been fertilized was upset by an accident. The basic pattern for part of the tissues of the embryo would be changed. As tissues grew according to the new pattern the final structure would have a new pattern and different properties. The shuffling of the order of the unit groups of protein chains may be a fundamental agency in evolution.

The determination of the order of the unit groups in a protein chain is difficult by the methods of organic chemistry, though new methods may make it easier in future. At present the X-ray technique is more powerful in the determination of the order of these units. A comparative study of the orders of the unit-groups in the proteins of various sorts of animals is becoming possible and will form a new science of anatomy. The old anatomy consisted of the study of the structure of living organisms and was confined by the limitations of the eye and microscope. The new anatomy will investigate the internal structure of the organs whose more external features only were, in the old anatomy, accessible to investigation. The technique of analysis by X-rays will revolutionize and vivify an ancient science that had become stagnant through the exhaustion of the old techniques, as the limits of profitable research by eye and microscope had been reached.

Certain fibres, such as wool, are notably elastic, while other fibres, such as natural silk, cotton, and artificial silk, are inelastic. A dry wool fibre can be stretched until it is one-third longer. In cold water it can be stretched until it is two-thirds longer, and in steam its length can be doubled. Yet in every case the wool fibre can be made to contract exactly to its original length. It resembles a strip of india-rubber. The photograph of stretched natural silk, artificial silk, and cotton fibres is substantially the same as that of the unstretched fibres. This indicates that there is no fundamental change in the structure of these fibres when they are stretched.

In contrast, the X-ray photographs of stretched and unstretched wool-fibres are quite different. This shows that the fundamental structures of stretched and unstretched wool fibres are different. The crystal units out of which they are built are of a different size. The X-ray picture of the stretched wool fibre has a very remarkable feature. It is substantially the same as that of the natural silk fibre. When the wool fibre is stretched to its elastic limit it is no longer elastic, and then it gives the same sort of picture as the inelastic natural silk fibre. Thus the internal crystalline structure of wool is changed by extension. When the stretched fibre is released, so that it returns to its original length, and again photographed, the new picture shows that the crystal structure has changed back again to its original form.

Somewhat similar phenomena appear in the examination of india-rubber. When it is unstretched it gives no X-ray structure picture, but when it is stretched an excellent picture is obtainable. Hence india-rubber is amorphous in the unstretched condition but crystallizes when it is stretched. The molecules of india-rubber, like the molecules of cellulose and proteins, consist of chains of groups of atoms. The fundamental group of the india-rubber chain is named isoprene and contains five atoms of carbon and eight of hydrogen. Unstretched india-rubber consists of a muddled mass of chains of isoprene molecules. They are coiled and twisted together without general structure. When the india-rubber is stretched the long isoprene chains are uncoiled and pulled out. This brings them into a more or less parallel alignment which is exhibited as a crystallization. India-rubber may be stretched nine times its length. This is a measure of the length and flexibility of the isoprene chains.

In the unstretched wool-fibre the molecules are not

lying in a muddled amorphous condition, but crimped. When the fibre is stretched the crimps are pulled out, and when they are released they return to their crimp structure (Fig. 30).

Natural silk and cotton are inelastic because their crystals are already in the fully extended form.

The elasticity of wool and hair fibres is due to the crimped structure of the constituent chains of molecules. These chains are normally coiled, but become

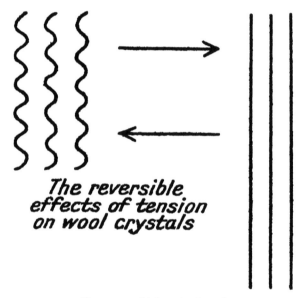

The reversible effects of tension on wool crystals

FIG. 30.   (After Astbury.)

uncoiled when stretched. The perfect elasticity of animal fibres is due to the perfect elasticity of their constituent molecular chains. This conclusion derived from X-ray studies probably applies also in the explanation of the elasticity of skin and muscle. The perfection of recovery is due to the perfect elasticity of the constituent molecular chains. Astbury explains the springiness of the chains in terms of electrical attractions.

The protein chain consists of groups of atoms. In parts of these groups there is a concentration of positive or negative electricity, though the sum of the positive and electrical charges for the group is zero. These irregularities in the distribution of electricity in a molecule may explain many of its qualities. They cause parts of the molecule to be attracted by other substances, when the remaining parts are unattracted because they are free from electrical irregularity. For

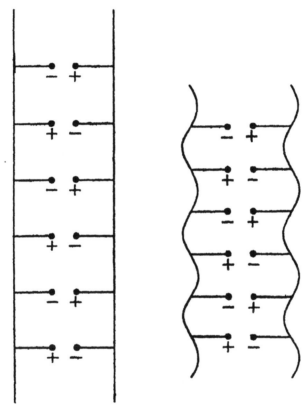

FIG. 31. (After Astbury.)

instance, some substances which do not mix with water nevertheless spread over the surface. One end of their molecule has an electrical charge which causes it to stick to the water while the rest of the molecule is indifferent.

Such a substance can cover a water surface with a pile of molecules, as a carpet consisting of a pile of fibres fixed on their ends. The atoms in the latter may have a tendency, also through electric charges, to stick together sideways. Indeed, this tendency may be sufficiently strong to enable the film to constrain the water as if it were a containing elastic skin. It is one of the qualities by which oil films calm rough water. The large total force exerted through the side-clinging tendency of myriads of oil molecules explains why troubled waters are calmed by pouring oil on them. Returning to Fig. 31, the chains in wool molecules are normally crimped because the electrical irregularities occur so that positive and negative charges are spaced alternately along the chain. The adjacent opposite charges attract each other and cause the chain to crimple or contract. When the chain is forcibly extended the attractions between the charges are overcome and the chain becomes as long as is possible without rupture. If it is extended further the major forces between the groups are overcome and the chain is broken. When the tension is removed the electrical attractions contract the chain into the original position. Hence a wool fibre is almost per-fectly elastic, as its elasticity depends on a perfectly elastic mechanism in the molecular structure. The elasticity of material produced by living organisms is one of the qualities which seem to make such products so different in nature from dead matter, yet it can be explained by molecular mechanisms.

The explanation of the setting of fibres by steaming and other treatment follows from the nature of their inner structure. When a wool fibre or a human hair is stretched to its elastic limit it becomes inelastic. So if they are stretched and the electrical charges which normally cause them to contract are put out of action by steaming or some other treatment, the wool or hair fibre will remain in any shape into which it has been bent while stretched. This is the explanation of the mechanism of "permanent waving" for ladies' hair. An X-ray photograph of a permanently waved hair shows that its crystals are of the shape found in inelastic fibres such as silk.

J. Thewlis has made an X-ray investigation of the apatite crystals which constitute the enamel of teeth. He has found human enamel contains two sets of fibres inclined at different angles to the surface of the tooth. In dogs' enamel the fibres are at right angles to the surface of the tooth. Three sorts of enamel can be distinguished. Susceptibility to decay in human teeth appears to be connected with two of these sorts of enamel, while immunity to decay appears to be connected with the third. X-ray investigation of the structure of the material of teeth is providing a new method of attacking the problem of dental decay.

Astbury and his colleagues have made some interesting researches on the structure of the cellulose in the walls of an elementary plant that lives in water, *Valonia ventricosa*, and grows as a single-celled hollow sphere whose diameter may reach almost an inch. The investigator prefers to work with structures produced by single cells because structures produced by groups of cells are more complicated and difficult to interpret. Hence single-celled organisms which produce structures of large size have a special investigatory value. An X-ray examination of the wall of *Valonia* shows it contains two sets of cellulose crystalline fibres crossed at a constant angle. The path of the fibre in the surface of the spherical cell-wall can be traced by taking X-ray photographs at neighbouring points and plotting the fibre directions on an inflated bladder as model. In this way, a map of the directions of the fibres is made. The results so far obtained seem to show the molecular structure of this natural cellulose *balloon* is built up in a spiral fashion on the same plan as the cellulose *fibres* ramie, cotton, etc. The cellulose crystals appear to be deposited out of the thin lining of protoplasm inside the cell. The X-ray photograph shows the crystals could not have grown by themselves but must have been laid down in certain directions as they are produced by the protoplasm. Hence cellulose structures appear to be an end-product arranged by the orientating mechanism of the protoplasm. The protoplasmic power of orientation is derived from the peculiarities of the fundamental scaffolding of living matter, the structure of protein molecules, which may be in the form of nets, chains, ladders, spirals, or other arrangements of atoms. The two sets of fibres in the wall of *Valonia* are built spirally. It is not yet known whether the two systems of spirals are continuations of each other, or whether they are distinct. If they are distinct then protoplasm must have the power of depositing cellulose rhythmically. The rhythmical properties of protoplasm may be derived from the elastic qualities of certain protein molecules, such as those of wool and other products of living bodies.

Thus the investigator is gaining some rational insight into the mechanism of the astonishing power of organization possessed by living matter. The problem of organization is the central problem of biology. It is yielding to the methods of scientific analysis, and there is now evidence that biological organization is a complicated expression of the peculiar structural properties of protein molecules. Simple mathematical calculations show molecules possessed of these properties may reasonably be built up into structures exhibiting the myriad subtleties and forms of living things.

CHAPTER VIII

HUMAN HEREDITY

I

THE vigour of a social class is exhibited by the tenacity with which it holds its opinions. The promulgation of the theory of evolution in the second half of the nineteenth century produced such intellectual uproar because the theory was both contrary to fantasies held by the new, powerful and creative capitalistic class and yet necessary to the philosophy of capitalism. The main interest of the nineteenth century was the creation of power industry. Its best energy was expended in the creation of a new sort of social organization involving the application of science to the methods of producing goods and to the achievement of ideals of life suitable to that type of social organization. Transport was developed by the application of steam partly in order to increase the volume of trade and also to amuse the rich rentiers produced by an increased volume of trade. The unabated advance of physical science during the nineteenth century was due to its connection with the machines of industry. Capitalistic development directed attention towards new ways of doing things in industry and science. It deflected attention from thought concerning the principles of social organization and of science. Attention to business prevented the leaders of capitalism from criticizing their own opinions. This is one explanation of why the nineteenth century, which achieved such greatness by discovering how to produce unlimited supplies of commodities, thus contributing more than any preceding century to human progress, could combine conspicuous energy of life with naïve opinions in philosophy and art. Britain was the leader of this wonderful nineteenth-century movement and preeminently exhibited the combination of industrial and scientific creation with philosophical and artistic poverty. The typical Lancashire cotton-mill owner belonged to sects which endowed the concepts of industrious sobriety and of individualism (expressed as the right and power of every man to have access to God and to make his own peace with Him) with particular sanctity. For their purposes the intellectual furniture of religion might, from lack of intrinsic importance, be borrowed from any period however incongruous. The mill-owner believed in his right to run his mill in his own way and to present his account of how he ran it to God personally. He accepted the Old Testament account of creation literally.

The Old Testament could not provide a philosophy adequate to a century which had surpassed all of its predecessors in creativity. The leaders of the nineteenth century required a philosophy more congenial to the principles underlying their activities. They could not continue indefinitely to ignore intellectual criticism and remain satisfied with a philosophy borrowed from primitive and extinct tribesmen. Their dogmatic belief in the sacredness of individualism had to be replaced by a rational theory. As the years of the nineteenth century passed away its leading classes became richer and more leisured. They had time to discover the naïveté of their philosophy. To explain their ascendancy their intellectual leaders wanted a theory more rational than belief in themselves as a people chosen but chosen without explanation. They found this theory in the principle of natural selection. The nineteenth century was far more interested in the principle of natural selection than in the general theory of evolution. This interest spread the popular fallacy that natural selection is not only the mechanism by which survival is decided, but also the creative force behind the panorama of evolution. Natural selection was supposed to be both the mechanism and the steam of the evolutionary engine, though obviously selection cannot be concerned with anything except selection from presented material. Natural selection has nothing to do with the origin of change, of variation. It merely operates on such changes and variations as occur in living organisms. It does not help to explain why these changes occur. The evolutionary theorists of the latter half of the nineteenth century believed in the importance of the principle of natural selection with an emotional violence from which Darwin himself was free. This conviction was an unconscious expression of class interest. The peculiar ferocity of the debates between the followers of T. H. Huxley and the flock of Bishop Wilberforce was because the former had the feeling that the latter were traitors to the dominant social class to which they both belonged. To the Huxleians Wilberforce and the group within the educated middle classes which he represented seemed to be enemies of truth and progress. They believed this not so much because it was true, but because the dominant social class to which they belonged endangered itself by continuing to accept an out-of-date philosophy.

This does not imply that the Huxleians were dishonest. Honesty is a function of the conscious processes. A man can believe with perfect honesty in the correctitude of theories which it is advantageous to himself and his class to accept. He is dishonest when he understands and will not admit that he holds certain opinions more through self and class interest than through intellectual conviction. The motives of persons and classes are usually clearer to others than to themselves. The correct theory of personal or class action is rarely given by the actors while they are in action. Hence the nature of the activities of the very active capitalistic classes of the nineteenth century were clearer to the intelligence of other contemporary classes than to their own. Members of the nobility and the working class were in a better position than the capitalists to understand capitalism. Lord Shaftesbury and Karl Marx understood capitalism better than the capitalists. Karl Marx could see capitalism from the standpoint of the two classes with which he mixed or had connections, the working class and the aristocracy. He had spent most of his life in contact with the working class, was born of a long line of rabbis and had married the daughter of an aristocrat. A class begins to see its early activities objectively only when they have occurred at a sufficiently distant date. At the present time many intellectual leaders of capitalism will admit some degree of correctitude in Marx's criticisms of capitalism. Their grandparents almost unanimously regarded them as insane ravings.

The acceptance of scientific theories is far from being dependent on their correctitude alone. A theory containing some elements of truth may be accepted wholly and with deep conviction if it happens to be consonant with the principles and ideals of a powerful class. There will be little difficulty in accepting without criticism the doubtful besides the sound parts of the theory. The biological theories developed during the nineteenth century were profoundly moulded by social influences. Theories of human heredity showed the effects of social influences as remarkably as the theory of the rôle of natural selection in biological evolution.

Among men of ability to whom the dominant classes of the nineteenth century provided means and hence the opportunity for research was Francis Galton. He founded the systematic study of human beings which provides the data for a science of social biology. Before biological theories concerning human beings can be developed exact biological knowledge of human beings must be collected. If knowledge is to take us far, it must be quantitative. Galton wished to investigate the nature of human beings as a biological problem and describe it with mathematical precision. There are two ways of arriving at exact mathematical descriptions of natural phenomena. The exact measurement and analysis of a few examples is one way, and the application of the mathematics of statistical analysis to a large number of approximate measurements is the other. Some phenomena are more amenable to the first method and other phenomena to the second. Many of the biological data of human beings are more amenable to the second, the statistical method, because lack of experimental control prevents the making of significant measurements of many properties of individual human beings. Galton sought particularly to discover exact theories concerning the biological nature of human beings by organizing the collection of vast quantities of measurements of human characteristics, analysing and summarizing them by the methods of mathematical statistics. He had been educated as a doctor and had travelled widely. This gave him a general knowledge of the biology of human beings, and of the data of anthropology. He had become accustomed to consider man as a biological organism and to compare the behaviour of different members of the species. He had been interested in meteorology, the science of weather, and the application of statistical analysis to meteorological observations for making weather forecasts. The publication of his cousin Charles Darwin's book on *The Origin of Species* quickened interest in biology, and Galton responded to the stimulus. When he left meteorology for social biology he took with him a knowledge of the methods of statistics.

Galton's success in inspiring the use of statistics in biology was such that the mathematical theory of statistics benefited from his researches and those of some of his followers. His pioneer work in mathematical biology as an independent branch of science was a first-rate contribution to culture and an excellent measure of the magnitude of his genius. The first enquiries in the new branch of science were incomplete; they were the beginning of researches which would continue for hundreds of years. The results of these brilliantly original early enquiries were necessarily based on the first meagre collections of data. Such researches, however brilliant, could not contain more than a few elements of certain knowledge. The degree of the conviction with which the results could be accepted depended therefore on theoretical dispositions of the intellectual public.

Some of the results of Galton's most tentative and incomplete investigations were accepted as thoroughly established knowledge by classes that wished to accept them as such. A distortion of biological theories often arises when they are applied to humanity. If they can be used to justify or excuse the activities of certain types of person or classes they may be accepted by these classes on very slight scientific evidence of their truth. Though this is true of general biological theories which apply to all living organisms without having more than an indirect application to human beings, it is even more true of biological theories that apply directly to human beings. Man's self-interest makes the task of the biological theoretician exceptionally difficult. An intelligence that functions excellently in a non-biological science may commit glaring errors in biology, and especially the biology of human nature, or social biology. An investigator in social biology cannot make important contributions without having a wide experience of life. He must be familiar with the views of the various social classes, preferably from personal contact, through political and business experience. Without such experience he has great difficulty in becoming in some degree conscious of the nature and origin of his own social prejudices, and cannot allow for their distorting influence on his conceptions of human behaviour. The social biologist must possess a wide experience and an extensive intellectual technique before he can make sound contributions to his subject. The leading research workers in social biology such as R. A. Fisher, J. B. S. Haldane and Lancelot Hogben are all distinguished by the range of their knowledge. In this chapter the arguments of Professor Hogben's admirable books *Nature and Nurture* and *Genetic Principles* will be closely followed.

An exact knowledge of the properties and mechanism of heredity in human beings must eventually become one of the bases of constructive social policy. What sorts of persons should be encouraged and what sorts discouraged in a progressive community? How should encouragement and discouragement be given?

Should unhealthy persons be discouraged from parenthood? Should victims of consumption have children, and should the mentally disordered be prevented from having children? None of these questions can be answered without exact knowledge of human heredity. And yet heredity is only one of the factors to be considered in answering them. Before advice or legislation can be given the influence of environment in producing admirable or unsatisfactory citizens must be measured. Are some persons intelligent, healthy and amiable because they have exceptionally good hereditary constitutions, or because they were born in a good environment and carefully reared? Or are admirable persons a joint product of good hereditary constitutions and good environments? If admirable types of men owe their qualities to hereditary constitution alone, improvements in the environment will not produce more of them. If this were true, policies of social amelioration could have no motive except human feeling.

The study of twins is an important method for investigating the interaction of heredity and environment in the development of human beings. Twins of the same sex and very similar characteristics are described as identical, and twins of the same or different sex with generally different characteristics are described as fraternal. Identical twins are produced by an accidental splitting of a single fertilized egg into halves that develop independently, and have the same hereditary constitutions. As they grow up they show differences that must be due to the effects of environment or nurture. Fraternal twins are due to the fertilization of two independent eggs about the same time and have different hereditary constitutions or natures. So their differences as compared with those of identical twins are partly due to differences of nurture and partly due to differences of hereditary constitution. Galton studied the characteristics of thirty-five pairs of identical twins and twenty pairs of fraternal twins in order to distinguish between the effects of nature and nurture in human beings. He described his results as follows:

"We may therefore broadly conclude that the only circumstance within the range of those by which persons of similar conditions of life are affected, that is, capable of producing a marked effect on the character of adults, is illness or some accident which causes physical infirmity. The impression that all this leaves on the mind is one of some wonder whether nurture can do anything at all beyond giving instruction in professional training. There is no escape from the conclusion that nature prevails enormously over nurture when the differences of nurture do not exceed what is commonly to be found among persons of the same rank of society and in the same country. . . ."

The results of Galton's admirable but necessarily preliminary researches into this and other aspects of human heredity have become the foundation for the policies of powerful societies. Many members of the successful classes in the competitive capitalism of the nineteenth century were predisposed to hear that they had been naturally selected for success in virtue of inborn qualities. They wished to go beyond even Galton and believe that the possession of comfortable homes and good education was of little advantage to them when in competition with persons of superior inborn qualities but of poor homes and education. Many successful members of a competitive social order were prepared to accept as thoroughly established biological knowledge any tentative evidence that the principle of 'every man for himself and the devil take the hindmost' was a suitable guide for the conduct of human behaviour. They wanted to be assured that their success as individuals was due to the inborn characteristics of themselves and that their success as a dominant class was due to its inborn biological superiority. They wanted scientific proofs that the unsuccessful owe their failure to their innate biological inferiority. If the dominant classes could be proved to be of superior biological quality they could feel that they were justified in establishing their dominance for all time. The acceptance of Galton's preliminary results as thoroughly established knowledge and their use as a basis for powerful agitation is an important example of the effect of social interests on the development of a branch of science.

Since the time of Galton a much larger body of information has accumulated. Larger numbers of twins have been examined and more precise ways of measuring their characteristics have been devised. The differences between identical twins are very small for some characteristics, such as height, weight or eye colour. In such cases fraternal twins are much more dissimilar, being no more alike than ordinary brothers and sisters. This is not true of all characteristics, even physical ones, such as the rate of breathing or the pulse. Several hundreds of pairs of twins have been examined by means of the intelligence tests to which we shall refer later. The average difference between identical twins is about half as great as the average difference between fraternal twins, and fraternal twins are decidedly more alike than ordinary brothers and sisters. This shows that even when children are brought up together in the same family the differences of environment to which they are exposed account for much of the difference of capacity between them. Such differences of environment must be very small compared with differences of environment to which children belonging to different families which in turn belong to different social classes or races are exposed.

If mental disorder, cancer, tuberculosis and other serious diseases are due to inheritable weaknesses of constitution, may not sterilization of persons suffering from them be a very effective method of improving the average health of the community? In Great Britain about 140,000 persons receive relief or treatment in public and private mental hospitals at an annual cost to the community of £10,000,000. About 1,000,000 persons, or about one in every forty, have received treatment at some time for mental disorder, often under the polite name of "severe nervous breakdown". Enquiry shows that nearly every family has some mental skeletons in its cupboard. The amount of mental disorder in the British population is said to be increasing, but the assertion should be received cautiously because old records of the incidence of mental disorder in the population are inaccurate. The amount of mental disorder in a population cannot be measured until mental disorder can be more exactly defined. Modern definitions are more precise than older ones, and the apparent increase of mental disorder or defect may be due merely to improved methods of diagnosis and public record. If mental disorders are inheritable and incurable, sterilization of sufferers might appear to be the only effective method of removing one sort of undesirable person from the community.

We now know that the mere stopping of breeding of certain types in a large population is not necessarily an effective and practicable method of reducing the extent of an inheritable disease. It depends on the way in which the disease is inherited and the frequency of its occurrence in the population. If a disease occurs with the same rarity and is transmitted according to the same rules as *albinism*, the sterilization of every new patient appearing during a thousand years would not reduce the frequency of its occurrence in the population by 50 per cent.   If the disease is inherited according to the rules obeyed in the transmission of a common form of night blindness, sterilization would remove it from the population within a generation.   Sterilization would be quite ineffective as a practicable method of reducing the incidence of some diseases in the population, but might be effective with others.   Nobody should entertain opinions on the value of sterilization without a knowledge of the laws of heredity and their operation in human populations.

All organisms which reproduce by a sexual mechanism have a mother and a father.   The mother supplies a comparatively large single cell which is the egg, and the father supplies a single cell named a sperm. When a new organism is produced it is made by combining one egg and one sperm to form one single cell.   This original single cell is the first stage of the new organism's existence.   It has the power of growth and multiplies by sub-division to form many more cells.   Skilful microscopical examination of this original cell shows that it possesses a nucleus which contains, as a rule though not always, an even number of bodies which readily adsorb certain dyes.   These bodies are named chromosomes (from the Greek words *khroma*: colour; and *somatikos*: body).   Microscopic inspection of the female egg and the male sperm cells shows that their nuclei contain chromosomes, but as a rule only half the number that appear in the cell which develops into an adult organism.   Evidently the new organism receives half its original supply of chromosomes from its mother and half from its father. When the original single cell sub-divides, the microscope shows that each of its chromosomes splits in half, so that the two new cells each have the same composition of chromosomes as the original cell. Hence the millions of cells in the adult organism contain as a rule exactly similar chromosome equipment; every cell in the body has as a rule the same number and same composition of chromosomes. The chromosomes are the carriers of hereditary potentialities. The father contributes his share of inheritable potentialities to the child through the chromosomes in his sperm, and the mother through those in her egg.

A number of consequences arise from the nature of this mechanism. In general the original cell contains two chromosomes influencing any one inheritable quality. For instance, one of the father's chromosomes may contain a factor or gene, as it is named, which would tend to produce a blue eye, while the corresponding chromosome from the mother has a factor or gene in it tending to produce a brown eye. Observation shows that human eyes are generally brown or blue, but not a mixture. Hence either the brown or the blue gene must take the precedence. Observation shows that when such a conflict occurs the brown gene wins. It is described as dominant over the blue gene, and the blue gene is described as recessive to the brown gene.

Evidently the cells of an offspring may contain a gene which is quiescent and does not exert an influence until it is paired with another of the same sort. Blue eyes occur in a purely blue- or brown-eyed family only when an offspring's cells happen to contain a pair of the sort of genes which produce blue eyes. A characteristic due to a recessive gene never appears in an offspring unless both parents have at least one gene for it. Neither parent may show the characteristic, in which case each has one gene for it; or one or both parents may have the characteristic, in which case one or both parents would each have a complete pair. When a disease is hereditary and due to a recessive gene it can appear in an offspring only when both parents carry the gene. If the disease is rare few people will carry even one gene for it, and the chances of persons who carry the gene meeting and having children are very small. This explains why sterilization is an ineffective method of obliterating rare inheritable diseases due to recessive genes.

The phenomena of dominance and recessiveness help to explain the biological ground for the human tradition against incest. Dangerous hereditary defects or diseases may tend to be due to recessive genes because natural selection would weed them out if they were due to dominant genes and appeared frequently in the population. Recessiveness protects a defect against natural selection. Since a dangerous condition due to recessive genes can only appear when both parents carry the recessive gene, the chances of this occurring are much greater if both come from the same family. Thus the marriages of closely-related persons tend to bring out recessive conditions. If the recessive condition is rare the chance that the two genes will come together in marriages of unrelated individuals is small. The marriage of close relations could bring out a rare recessive condition in a large number of the members of their family, when marriage with outsiders would rarely produce one example of the condition. The human species may or may not contain a large number of rare recessive genes which cause dangerous diseases or conditions, but if it does, incest would make the appearance of these conditions very much more frequent.

By one of the most fortunate circumstances in the realm of natural phenomena the mechanism of heredity operates according to simple mathematical laws. This is one of the happy consequences of its particulate nature. As inheritable characters are due to the positions and assortments of definite units their distribution among offspring may be calculated. In simple relationships the calculations are very simple. For instance, if one parent has two genes for brown eyes and the other has two genes for blue eyes, all of the children will have brown eyes. If one parent has one gene for brown and one for blue, and the other has two for blue, half of the children will have brown eyes and half blue. If both parents each have two genes for blue, all the children will be blue-eyed. If each parent has one gene for brown and one for blue, one quarter of the children will have blue eyes and three-quarters will have brown eyes. If one parent has two genes for brown all of the children will have brown eyes, whether or not the other parent has one or both genes for blue.

These simple numerical ratios follow immediately from the nature of the chromosome mechanism. For more complicated relationships and distribution of genes the calculations are not simple. The elucidation of the mathematical theory of the distribution of inheritable characteristics in a population under various circumstances is being pursued actively at the present time, and is one of the most important developments in biology. The leaders in this field of research include F. Bernstein, J. B. S. Haldane, L. Hogben, H. S. Jennings and Sewell Wright.

The new investigations of the mathematics of inheritance show that parent-child and brother-sister unions will bring forth the largest, and equal, numbers of rare recessive conditions. An uncle-or-aunt-with-niece-or-nephew union, or a half-brother-with-half-sister union or a grandchild with grandparent union are equally effective, but only a little more than half as effective as the parent-child union. First cousins are rather more than one quarter as effective and second-cousins rather more than one-sixteenth as effective. The strength of the traditional opposition to marriages of relatives is roughly proportional to the closeness of the relationship and this is proportional to the effectiveness with which the marriages may bring out recessive conditions. Inbreeding brings out recessives. If conditions are good inbreeding should improve the species, but if they are bad it should worsen it. Though at present the knowledge of the nature and number of recessive conditions in human beings is insufficient to say whether one alternative occurs more often than the other. The average person would certainly be frightened of inbreeding if he understood the extent to which it can bring out hereditary diseases due to recessive genes.

As inbreeding brings out rare recessive genes, their appearance indicates that their possessor is probably a child of closely-related parents. This is confirmed by the well-known observation that persons suffering from rare hereditary diseases usually belong to families in which consanguineous marriages have been exceptionally frequent. One type of deaf-mutism is a recessive condition. According to Lenz its incidence is four and a half times as great among the Jewish than among the Gentile population in Berlin. This is compatible with the higher percentage of marriages between relatives in the Jewish community. There are other kinds of deaf-mutism which may result from attacks of meningitis, scarlet fever, syphilis and mumps in early childhood. The interpretation of the evidence for the inheritability of this disease must be made with the caution that is necessary in all considerations of the rôle of heredity in human beings. Through its effect in bringing out rare conditions inbreeding produces more variety in the characteristics of offspring. The children of inbred families exhibit an exceptionally wide range in their measurable characteristics.

The chromosome mechanism of heredity produces other interesting effects besides the mathematical system of the distribution of inheritable characteristics. The cells of organisms usually have an even number of chromosomes. For instance, the cells of the vinegar fly *Drosophila* contain eight chromosomes in four pairs. The genes which determine the various inheritable characteristics of the organism are situated in the various chromosomes. The chromosomes are in fact groups of genes. Thus particular sets of characteristics are inherited in groups because the chromosome which contains the gene for one of the characteristics also contains the genes for others. As the cells of the *Drosophila* fly contain four pairs of chromosomes, its hereditary varieties would be expected to fall into four groups. A very thorough investigation of the heredity of this insect has shown that all of its recognizable hereditary varieties, about 400, may actually be sorted into four such groups. The phenomenon of linkage arising from the limited number of chromosomes has particularly interesting effects in connection with the determination of sex. As with many other characteristics, the sex of an organism is partly determined by a different equipment of genes. In general each cell will contain two chromosomes bearing genes specially concerned with sex; one from the mother and one from the father. These chromosomes may contain genes determining other characteristics, so the distribution of the latter will be connected with the sex of the organism. The chromosome mechanism in the cells of human beings exhibits a peculiarity, as the cells of women contain forty-eight chromosomes in twenty-four complete pairs, whereas the cells of men also contain forty-eight, including twenty-three complete pairs, and one pair of which one member is a mere vestige. This vestigial chromosome of the human male may be regarded as a dummy. The sex of a male human being is thus determined by the absence of a complete complementary chromosome in one of his twenty-four pairs.

The transmission of the disease known as haemophilia is an example of the way in which genes housed in the sex chromosome behave. Patients with this disease are liable to die of bleeding from a slight cut because their blood does not coagulate. The characteristic feature of such diseases is that we can usually find another patient among the ancestors, if we trace backwards along the female line, but never by tracing through the father's pedigree. Another feature is that when such diseases are rare practically no female patients exist. In the case of haemophilia female cases never occur, but this is not for the same reason as the one which explains the comparative rarity of female patients in other sex-linked conditions. It seems that the gene cannot exert its effect upon the blood in the kind of environment which the female body supplies. This suppression of its influence may be due to a secretion of the ovary, such as the ovarian hormone named oestrin which is connected with the monthly period. Recently some success has been obtained in curing males by ingrafting ovarian tissue or injecting oestrin. The blood of haemophiliacs can thus be made to clot normally.

This discovery shows once more the necessity for caution in advocating sterilization as a measure against inheritable diseases. It is quite possible that even so serious a disease as haemophilia may be rendered harmless through the discovery of a satisfactory treatment. A distasteful inheritable disease, alkaptonuria, is characterized by the darkening of the patient's urine after it has been exposed to the air. The darkening is due to the oxidation of a substance named homogentsic acid. Normal persons possess a chemical enzyme which decomposes homogentsic acid. Biochemists may one day discover how to manufacture this enzyme and provide alkaptonuriacs with a satisfactory treatment.

The control of inheritable diseases does not necessarily involve discouraging sufferers from breeding. Hogben has remarked that two hundred years ago inheritable differences of resistance among Englishmen probably had a large part in deciding whether a particular Englishman would die or escape in a smallpox epidemic. To-day such inheritable differences have little importance in connection with small-pox. We can raise the resistance when it is low by vaccination and we can shield an individual with low resistance from exposure to infection. In the course of thousands of years a European community might evolve a high degree of immunity to small-pox through selective elimination of the non-resistant. The African peoples have probably evolved their high immunity to malaria through such a process, but, as Hogben writes, thanks to human inventiveness, we have not had to wait several thousand years to get rid of small-pox. Inheritable diseases may often be treated more effectively by controlling the environment, the diet, and so forth, than by controlling breeding.

Like haemophilia, colour-blindness is another disease which is hereditary and sex-linked. About 4 per cent of the male population suffer from it, and only 0·4 per cent of the female population. Sex-linked recessive conditions are recognized more easily than non-sex-linked because of the pronounced restriction of the condition to one sex. This is probably the explanation why most of the known rare recessive conditions in human populations are sex-linked. The discovery of the rare non-sex-linked conditions is much more difficult. The phenomena of sex-linkage in inheritance makes selection or sterlization a very much more effective agent for removing recessive conditions from the population. If all the males suffering from a sex-linked recessive condition are prevented from breeding the number of sufferers in the population may be reduced by 50 per cent within a generation. Sex-linkage also profoundly affects the results of inbreeding. Under certain circumstances, i.e., when sex-linked genes are involved, a union with a paternal uncle may involve consequences of the same order of magnitude as marrying a sister, whereas a union with a paternal aunt may involve no deviation from mating at random or marrying a deceased wife's sister. According to British law it is illegal for a man to marry his aunt, but it is not a criminal offence for aunt and nephew to produce an illegitimate child. It is a criminal offence for a brother and sister to cohabit, whether they produce offspring or not. Thus British law on incest is only partially compatible with the biology of inbreeding. On this matter Hogben remarks with some justice that: "The constitution of Great Britain endows bishops with special prerogatives to deal with such issues. It does not extend the same prerogatives to biologists."

The exact determination of the rules of inheritance of the less abnormal and normal inheritable characteristics in human beings is much more difficult than when we confine ourselves to striking abnormal variations. Before the rules of inheritance of a characteristic can be determined the characteristic must be exactly defined. The biologist can now aim at making a complete catalogue of all the inheritable characteristics of man and of the positions of their genes on the various human chromosomes. A few years ago this seemed an impossible dream, but the remarkable investigations of Bernstein on the inheritance of the blood groups have made the prospect not only admissible but hopeful.

The ancient medical technique of blood-transfusion is well known. Persons suffering from loss of blood by wounds or disease may often benefit from an infusion of blood from a volunteer. The technique is very ancient, as there is evidence that prehistoric men practised it. The prehistoric belief in the life-giving properties of blood would easily suggest that a man could receive new strength from an infusion. When transfusions are made under perfect surgical conditions the blood of the recipient is sometimes curdled or agglutinated by the infused blood. The patient's circulation becomes impeded by the thickened blood and gradually stopped, so that death occurs. The curdling effect of the two bloods may be demonstrated in a test-tube. It is due to the reaction of substances in the serum with substances in the corpuscles. The bloods of two persons may contain different varieties of both sorts of substances. Human beings may be divided into four groups according to the mutual curdling properties of their bloods. The first group consists of persons whose blood corpuscles are not agglutinated by the serum of the bloods in the other groups and whose serum will agglutinate the corpuscles in the other groups. The fourth group consists of persons whose blood corpuscles are agglutinated by the sera from the other groups.

The second group consists of persons whose blood corpuscles are agglutinated by the sera from groups one and three only, and whose serum will agglutinate the blood corpuscles in groups two and four. These relations may be explained on the assumption that there are two varieties of reacting substances in blood corpuscles and two varieties of reacting substances in serum, and that these may be assorted together in four different ways.

The distribution of these four substances in the human species obeys the laws of heredity very exactly, and may allow the chromosome position of their own genes and the genes of linked characteristics to be determined. Further research may reveal the existence of many other substances of a similar sort. The determination of the group to which a patient belongs enables the surgeon to choose as the blood-giver a volunteer who belongs to a group whose blood will not curdle the patient's. This has materially reduced the percentage of deaths occurring in transfusion operations. Besides the great contributions to surgery and the theory of human heredity, the study of the blood groups has provided an important new sort of legal evidence of paternity and promises important contributions to anthropology. As the agglutinating properties of a blood are inheritable, evidence concerning the paternity of its possessor may in certain circumstances be deduced from them. In Germany, Austria and the U.S.S.R. such evidence is now legally recognized and has been acted upon. If the group of a child and the group of one parent is known, the other parent will not belong to one or two of the other groups. Hence, if a woman has an illegitimate child and asserts that a certain man is the father, the determination of the blood group to which the man belongs may show that his blood does not belong to the group or groups to which that of the father of the child must belong. The woman's assertion would then be proved incorrect and the man would be saved from unjust exactions on behalf of the child.

The study of the distribution of the blood groups among the races of mankind offers interesting evidence of racial relationships. Among normal Americans about 45 per cent belong to group one, 41 per cent to group two, 10 per cent to group three, and 4 per cent to group four. The percentages in populations in various parts of the world vary, and the gradual changes in the percentages may be followed with the change of topography. Thus the world has as it were been colonized by blood groups and the trail of these groups promises to show unsuspected relationships between the various coloured races and racial types. These researches may help to provide the evidence that will destroy the inter-racial prejudices leading to such terrible calamities in many parts of the world. Bernstein concludes from his study of the topographical distribution of the blood groups that there was an original group not now found in a pure form in any part of the world. The Filipinos and American Indians are the least changed from this original group. A second race arose from the original group in Asia. The maximum percentage of groups three and four is found among the Japanese. Another race is supposed to have arisen in Europe and travelled into India and Manchuria. Similar agglutination phenomena are encountered among the apes. The gibbons and chimpanzees belong to group two, while the orang-outangs belong to groups two and three. Man and the modern apes may have derived their blood groups from blood groups which already existed in that common ancestral stock from which both are probably descended.

A classification of human races according to their blood groupings disagrees with other classifications, such as those according to colour of the skin. This is an additional indication of the mixed pedigree of existing human races. A race of biological organisms is said to be hereditarily pure when the characteristics of every male are similar, and those of every female are similar. For instance, every member of the race might have fair hair, blue eyes and be at least six feet tall. However the members of the race mated, all their offspring would have these three qualities, if the race were hereditarily pure in these qualities. Purity of race for a given characteristic implies in general that their chromosomes always contain pairs of the sorts of genes determining the appearance of the fair hair, blue eyes and tall stature. Stock-breeders have bred races of cattle, rats, fowl and other animals which possess a high degree of hereditary purity. They have accomplished this by a rigid control of animal breeding. Offspring with unwanted characteristics have not been allowed to breed, and the animals for mating have been carefully selected. After generations of severe control relatively pure breeds, such as Jersey cows, have been produced.

Stock-breeders know from experience that the offspring of crosses between two relatively pure races of cows are usually inferior to the parents in the sort of characteristics which gave them value as domesticated animals. Certain sociologists have argued by analogy that as the crossing of pure races of cattle is disadvantageous the crossing of human races is disadvantageous. This argument has at the present time sociological importance because parties in possession of power believe it is true. Closer scrutiny shows how little value it possesses. It assumes the human races are hereditarily pure; in fact, that they are races in the biological sense. The biological evidence shows they are thoroughly mixed. It assumes desirable human characteristics are as easily definable as a high milk yield in a milch cow. It assumes human beings do not mate at random. It assumes their marriages are as selectively controlled by external agencies as the marriages of pedigree cattle. The biological evidence against the desirability of interbreeding between white, black, brown, and yellow human beings is slight. The objections to interbreeding at the present time are mainly due to dogmatic opinion based on human colour and what is named race-prejudice. It should be understood that the word 'race' when applied to Negroes, Mongolians, Jews or Caucasians is not a biological term, but a word for describing groups of persons who have a few characteristics in common.

There are no serious biological objections against human half-castes. Such biological objections may be discovered in future, but they are as yet unknown. The valid objections are utilitarian and sociological. The existing prejudice against half-castes makes their life miserable. Persons are recommended to refrain from producing half-castes because their offspring will have a socially unhappy life, until prejudice against human colour mixtures is dissipated. The degree of colour prejudice is a sociological not a biological phenomenon. The differences in attitude towards coloured persons in England, France, America, and Russia suggest objections to coloured persons is a social prejudice without a physiological foundation.

These considerations show also that theories which ascribe the rise and fall of civilizations to the genetical constitution of human tribes should be received with great reservation. Hogben has written that:

"the development of early civilizations by the coloured races when the Nordic peoples were still barbarians, does not compel us to believe in the inferiority of the Nordic people. Conversely, the achievements of more backward peoples in the present era do not compel us to assume they are incapable of assimilating our own type of social organization. . . . At a time when we hear so much of the superiority of the Nordic race it may be well to bear in mind the views of those who were preparing the ground for the cultural development of Northern Europe when our own forbears were little better than barbarians. A Moorish savant, Said of Toledo, describing our ancestors beyond the Pyrenees, observed that they 'are of cold temperament and never reach maturity; they are of great stature and of a white colour. But they lack all sharpness of wit and penetration of intellect.' This was at a time when a few priests in Northern Europe could read or write and when washing the body was still considered a heathen custom, dangerous to the believer, a belief that lingered on to the time when Philip II of Spain authorized the destruction of all the public baths left by the Moors. The Moorish scholars of Toledo, Cordova, and Seville were writing treatises on spherical trigonometry when the mathematical syllabus of the Nordic University of Oxford stopped abruptly at the fifth proposition of the first book of Euclid."

The determination of the respective rôles of nature and nurture in the formation of human adults is of fundamental importance for the construction of social policies. It is difficult because it is a complicated problem and because the human mind has special difficulties in approaching problems involving itself. The investigator can only study those factors of which he is aware. When he considers human problems he inevitably considers his own experience. One of the most obvious features in his memory of experience is that, as a rule, extremely little concerning the first four years of life can be remembered. Hence the human observer starts with a notion that he began to be influenced by the environment mainly after he was four years old. The Freudian school of psychologists have done much to dissolve this notion. They have discovered some good evidence and a large quantity of less good evidence that infantile experiences have an important part in the formation of the characteristics which appear later in the adult.

No adequate notion of the rôle of environment in the production of characteristics can be obtained without a study of the influence of environment in the period between conception and the fifth year, and particularly during the nine months of pre-natal existence. Reliable evidence of the influence of pre-natal environment exists. Dahlberg has given a table of the percentages of identical and fraternal twins born in Danish and French populations.

COUNTRY.	Age of Mothers.						
	15–20	20–25	25–30	30–35	35–40	40–45	45–50
Denmark (1896–1910):							
Identical . . .	0·42	0·36	0·35	0·41	0·47	0·40	0·31
Fraternal . . .	0·36	0·57	0·87	1·18	1·57	1·25	0·30
France (1907–10):							
Identical . . .	0·31	0·31	0·34	0·38	0·38	0·36	0·35
Fraternal . . .	0·25	0·45	0·73	1·06	1·44	1·12	0·34

The figures show that the chances of a mother bearing identical twins are about the same at all ages, whereas the chances of bearing fraternal twins are five times as great at the age of about thirty-five as at eighteen. There is some evidence, though of dubious value, that the chances for twins increase with the number of pregnancies, independent of the mother's age, being three times as great between the eighth and tenth as between the second and fourth pregnancies.

Apparently the state of the mother's womb has an important influence on the chances that she shall bear fraternal twins, but not on the chances that she shall bear identical twins. The womb environment is different for each member of a pair of twins. If it were not, identical twins would always be born in the same state of health. In fact one member of a pair of identical twins is sometimes born dead, while the other is alive. The genetical similarity of identical twins is confirmed by the evidence that the probability that one twin will survive if the other dies is less than for fraternal twins. Other distinct evidence of the influence of pre-natal environment is given by studies of Mongolism. Offspring born with this condition may occur in a family of which the other children are normal. It is a form of idiocy associated with anatomical peculiarities in the eyelids, toes, tongue and elsewhere. A study of the data collected by Dr. Shrubsall and by Dr. Lionel Penrose shows that Mongoloid children are borne by mothers of forty years with much greater frequency than by mothers of twenty years. Also the chances of a Mongol appearing are much greater after a large number of children have already been borne. If child-bearing were confined to the period between twenty and thirty years of age, the occurrence could be very much reduced. While hereditary constitution may have some influence in determining whether a mother *can* bear a Mongol, there is no doubt that pre-natal environment has a major influence in deciding whether she *will*. The numerical order of birth in a family is an important formative factor. According to the mechanism of heredity, the chances that a tenth child should have as healthy a genetic constitution as a second child are the same. Hence differences in the health of children connected with the order in which they are born are probably due to the faulty functioning of an overworked womb. There is evidence that the less extreme forms of mental defect occur more frequently in first-born than in later children. This suggests that the difficulties of a first birth tend to damage a child's constitution. When a womb has had moderate practice it works more efficiently. Karl Pearson writes that "the growth of the first child is hampered by conditions which exist to a far less extent for the following births; but these conditions will be much harder for the first-born when its mother is forty than when its mother is twenty-five".

The extent to which differences of environment and differences of hereditary constitution determine the intelligence of human beings are of great sociological importance. The definition of intelligence is difficult. The most satisfactory is given by tests of the sort introduced by Binet. Intelligence tests are often supposed to measure innate differences of intelligence independent of nurture and education, and consequently hereditary. Different schools of investigations differ on this point. Miss Burks of California concluded from a study of orphans that home environment contributes about 17 per cent of the variance in intelligence, parental influence about 33 per cent; the total contribution of heredity being about 75 or 80 per cent. Freeman, Holzinger and Mitchell of Chicago tested the intelligence of orphans before and after entering orphanages and foster homes in early childhood. They found that their intelligence had increased after several years' residence. They found that brothers and sisters reared in different foster-homes have a lower coefficient of resemblance in intelligence than brothers and sisters normally reared in their parental family. They also found that children sent to poor foster-homes performed less well in the tests than children sent to better foster-homes. Two unrelated children reared in the same home were found to resemble each other in intelligence. Enquiry showed that a large percentage of the parents of the orphans were mentally defective, yet the average mentality of the foster-children was little below normal.

Studies of the distribution of intelligence in the population as measured by intelligence tests show that it does not conform to any simple rules of inheritance. This does not necessarily imply that intelligence is not in part or even in the whole determined by heredity, but it implies that the mechanism of the inheritance of intelligence must be determined by many genes simultaneously. J. B. S. Haldane has shown that single characteristics determined by a definite group of recessive genes are much less susceptible to modification by selection than characteristics determined by single genes. If intelligence is determined by heredity selective methods could conceivably be used to increase the level of intelligence in the population, but the effects of these methods could not be foreseen until very much more is known of the mechanism of the inheritance of intelligence than is known at present.

The present obscurity concerning the respective rôles of heredity and environment in the determination of intelligence makes the collection of valid scientific evidence for the desirability of selective treatment in mental defect, education and social service difficult. There may be real hereditary differences of intelligence between different human races and between different social classes, but the dubiety of the scientific quality of most of the existing evidence for these differences cannot be exaggerated. Policies for the social control or sterilization of persons of moderate degrees of mental defect cannot at present be founded on solid scientific knowledge. The social biologist can agree to recommend the legalization of sterilization of persons suffering from some but not necessarily all rare and extreme forms of definitely hereditary disorder. The advances of medical research may alleviate even extreme hereditary disorders. For instance, the discovery of a drug allied to oestrin might enable a genius suffering from haemophily, or continuous bleeding, to live a comfortable life.

The sterilization of persons suffering from mild mental defect cannot be advocated until their affliction is conclusively proved to be due to particular forms of heredity. Family pedigrees do not necessarily provide indubitable evidence for the inheritance of mental defect. T. H. Morgan has remarked that the prevalence of mental defect in a pedigree may indicate merely that a family once in a poor environment has been compelled by the environment to stay there. J. B. S. Haldane has commented on the poor scientific quality of the evidence for mental defect in family pedigrees. H. H. Goddard has been prominent in the study of the pedigrees of families containing mental defect and widely quoted. Haldane quotes an example of Goddard's method of determining mental defect. He is describing the parents of a feeble-minded child.

> "Both parents are feeble-minded. The father is very high grade, so that for a considerable time we were much in doubt as to how to classify him. His feeble-mindedness takes the form which makes him noted as being peculiar. He is ignorant, lives alone, but is a good workman, sober, honest and industrious."

Haldane remarks:

> "A good workman, sober, honest, and industrious, but not good enough for Dr. Goddard. It seems to me perfectly monstrous that we should be asked to interfere with our fellows on evidence of this kind."

While there is good scientific evidence for the desirability of the compulsory sterilization of a very few persons suffering from certain extreme and very definite disorders, and the voluntary sterilization of persons suffering from some other extreme and definite disorders, there is no good scientific evidence for the desirability of direct or indirect sterilization of large social groups or classes. Haldane writes that the danger of multiplication of the mentally defective is real, but unimportant compared with the danger that governments should sterilize, segregate or starve persons defined as mentally defective because they belong to the Opposition. For instance, some British publicists have attacked the provision of poor relief and transitional benefit to the unemployed on the ground that this class is on the whole congenitally inferior. Haldane writes:

> "To my mind the attempt to justify such measures on biological grounds is a prostitution of science, far more serious than the manufacture of high explosives, bombing aeroplanes or poisonous gases. We biologists cannot prevent statesmen from doing these things, but we can most emphatically protest against their being done in the name of biology, and in countries where speech is still free we can warn the public against this misuse of our science."

The study of the rôle of heredity in human beings is hindered by the absence of uniform social conditions. Hopkins has remarked that the recent advances of biological knowledge have owed much to the more thorough use of the method of controlled experimentation, that is, the making of experiments on animals which live in environments virtually identical except for one or more distinct, known and measured differences. If all human beings had equal incomes, equal and similar food rations, similar houses all facing south, etc., the researches of the social biologist would be easier. At present he has to depend on the application of refined methods of analysis to adventitious features of human society, such as the comparison of the intelligences of identical twins that happen to live in different homes and environments, or of children of different ancestry that happen to live in the same home, etc. The satisfactory conduct of such analyses demands high technical equipment, and one can be sure that persons without this technical equipment cannot have first-hand scientific acquaintance with the established biological knowledge concerning human heredity.

Hogben concludes that the present state of the biological knowledge of human heredity provides useful guides for the treatment of some conditions definitely of a medical nature, and generally accepted as falling within the sphere of a medical doctor's control. It has at present little to offer towards the determination of sociological qualities, such as the comparative congenital intelligence of human races and classes. Restriction of the amount of educational facilities and social services given cannot be justified on biological grounds. There is no sound biological evidence for ascribing the rise and fall of civilizations to the hereditary constitution of particular peoples. He contends that sociology may learn more instructive lessons from the organizing mechanisms of the central nervous system than from human heredity. The present difficulties of humanity are due to social disorder, and the cure of social disorder will with much greater probability be found in the learning of new social habits than in producing new hereditary constitutions. Human behaviour can probably be modified much more easily than human hereditary constitution.

Those inclined to impatience because 10 million pounds are spent every year in Britain on the support of persons parasitic through mental defect and some millions of pounds are wasted on students unsuited to the sort of education given them, might reflect that over 100 million pounds are spent every year on armaments, and much more on social parasites who receive incomes grossly disproportionate to their contribution to society. In Hogben's words: "the selfishness, apathy and prejudice which prevents intellectually gifted people from understanding the character of the present crisis in civilization is a far greater menace to the survival of culture than the prevalence of mental defect in the technical sense of the term".

CHAPTER IX

PERNICIOUS ANAEMIA

I

IN 1926 G. R. Minot and W. P. Murphy of Harvard University described the effects of feeding patients suffering from pernicious anaemia with a special diet rich in liver. They found that the health of the majority of the patients, who would in the previous course have died within three years, was improved, and in many cases was restored sufficiently to allow a return to the occupations of ordinary work and business, by the daily consumption of upwards of half a pound of liver. As liver can be eaten by the patient and need not be injected by any special technique, the treatment is very practicable. Its practicability compares favourably with that of insulin for diabetes, as insulin must be injected and the continued puncturing of the skin may cause soreness and a psychological revulsion against the treatment by 'getting on the patient's nerves'. Many authorities consequently regard the liver treatment of pernicious anaemia as the chief of recent discoveries in that branch of medicine concerned with the treatment of diseases by remedial agents. For these researches on the cause and treatment of pernicious anaemia Minot, Murphy and Whipple have been awarded the Nobel prize for medicine in 1934.

A full account for the specialist of the history, nature and treatment of the disease has been given by L. S. P. Davidson and G. L. Gulland in their excellent book *Pernicious Anaemia.* The narration in this chapter follows their account closely.

The earliest known recognition of pernicious anaemia as a definite disease appears to have been made in 1824 by a Scottish physician, Dr. J. S. Combe. In the *Transactions of the Medico-Chirurgical Society of Edinburgh* he gave a description of a patient suffering from a severe anaemia which evidently was pernicious anaemia. Combe wrote:

> "The case now recorded appears to me entitled to still further attention as exhibiting a well-marked instance of a very peculiar disease, which has excited little attention among medical men, and which has been altogether overlooked by any British author with whose writings I am acquainted. Unfortunately, however, such is the allowable diversity of opinion on most medical subjects that it is very possible the following case may be viewed in different lights, and receive different appellations; and while some may be disposed to regard the peculiar characteristic from which it derives its denomination of anaemia as constituting a morbid state *sui generis* (a disease in itself) others may consider the defect of the red circulating mass as an accidental and occasional circumstance, denoting some peculiar change in the assimilative powers, the primary stages of which we have been unable to detect. Doubtful myself which of these opinions may be the most correct, I shall do little more than state correctly the phenomena of the case and minutely the appearances presented on dissection. One remark only I may at present offer, that if any train of symptoms may be allowed to constitute anaemia a generic disease the following may be considered an example of it in its most idiopathic (pure and not occasioned by another disease) form."

This passage shows Combe believed he was describing the effects of a definite disease which might be due to some disturbance of the assimilative mechanism, that is, to some disturbance of the body's process of assimilating material from food into its constitution.

He continues:

"The case was a man, aged forty-seven, who had been born and had spent the greater part of his life in the country, where his duties were neither laborious nor unhealthy; who had led a regular and temperate life, and had enjoyed perfect health since childhood, and had never lost any blood. I was much struck by his peculiar appearance. He exactly resembled a person just recovering from an attack of syncope (fainting); his face, lips and the whole surface were of a deadly pale colour; the whites of the eyes bluish; his motions and speech languid; he complained much of weakness; his respiration, free when at rest, became hurried on the slightest exertion; pulse 80 and feeble; inner part of the lips and fauces (throat) nearly as colourless as the surface; bowels very irregular, generally lax, his stools very dark and fetid; urine reported to be copious and very pale; appetite unimpaired, of late his stomach has rejected almost every kind of food; constant thirst; he has no pain referable to any part, and a minute examination could not detect any structural derangement of any organ.

"It was only about two months ago since he began to complain, but not until his friends had observed his altered complexion; he then lost strength and said his head troubled him. Of this last symptom he has no distinct recollection. His feet became oedematous (swollen, dropsical) and his appetite failed him.

"My attention was drawn to the skin, which was of waxen colour, soft and delicate, the cellular tissue about the eyes and breast slightly distended with watery effusion. The pulse was feeble, and easily excited by any emotion. A very minute examination of the case, and a careful consideration of its history, scarcely solved the nature of the affection, and its long continuance and inveteracy rendered our prognosis (the forecast of the course of the disease) much more doubtful.

"He died six months after being first seen, from aggravation of all his symptoms, extension of the oedema (swelling) to face and upper extremities, and evident marks of effusion into the chest. He died with all the symptoms usually attendant on hydrothorax (dropsy of the chest). At first the treatment seemed to check the disease, but latterly the stomach and bowels became so irritable as scarcely to admit of any medicine and only of the mildest diet."

At the post-mortem Combe noted that the fat beneath the skin was scanty,

"of a pale yellow colour, and semi-fluid. Not a drop of blood escaped on dividing the scalp. The heart when cut into was of a pale colour, and did not tinge linen when rubbed on it; it appeared like flesh macerated many days in water. The right ventricle contained a pale coagulum. The left side was wholly empty. There was a considerable moisture bedewing the viscera of the abdomen. The liver was of its proper size and structure, but of a light brown colour. The spleen was the only viscus of its usual colour; it was very soft, and its contents, on pressure being applied, turned out like a sac. The kidneys were nearly bloodless; the pancreas of a pale reddish hue. The stomach and intestines were perfectly sound, thin; showing no vessels, and transparent. The muscular substance throughout the body was like that of the heart, very pale, and exuded no blood, but only a pale serum, when cut into. The arteries were empty, and so were the jugular, femoral and humoral veins."

Combe's ably observed and expressed description was followed by other descriptions of single cases, but received no special attention. In 1849 and 1855 Dr. Thomas Addison described examples of the special form of anaemia known by his name. He asserted this form of anaemia was a definite clinical entity and was not a secondary anaemia due to the disturbance of the body by another disease, such as cancer. In 1867 and 1872 Professor Biermer of Zürich independently concluded that some anaemias were definite and not derivative clinical entities. In addition to the points noted by Addison he observed that the patients suffering from what he named progressive pernicious anaemia had small broken blood-vessels, particularly in the retina of the eye. Biermer included some secondary anaemias within his definition of pernicious anaemia, as he mentioned that it was found in certain cases of cancer and infestation, with intestinal parasites. He believed, too, that it might be produced by bad living conditions, as it appeared particularly among the poor, and was associated with unsuitable feeding, and chronic diarrhoea. Biermer regarded pernicious anaemia as a clinical entity produced by many different causes, and his opinion was supported by Ehrlich and others.

William Hunter was the first to recognize that the stomach, the intestines and the nervous system, besides the blood system, had an important rôle in the disease. Hunter noted that the stomach and intestines in every case were septic and contained an abnormal quantity of microscopic organisms. He observed the deposits of iron in the liver, and the infection of the stomach and intestines by organisms from sepsis in the mouth. He believed that studies of the state of the blood gave little information, and that the disease was due to sepsis which destroyed blood corpuscles in the portal veins, which are blood channels connecting the intestines with the liver.

The investigation of the state of the blood received a powerful impetus from Ehrlich's invention of the use of aniline dyes for staining blood films. This allowed the more elusive types of blood cells to be studied much more profoundly, and also the process of the formation of the blood in the marrow of the bones. As a result Ehrlich concluded that the primary site of the disease was the bone marrow, where defective manufacture of the blood was occurring.

A failure of the stomach to secrete hydrochloric acid has long been recognized as a prominent feature of pernicious anaemia. It is only within the last three years, owing to the brilliant researches of Castle of Boston, that the essential cause of pernicious anaemia has been discovered. Castle found that when beef steak and normal human gastric juice was incubated for two hours and fed to a patient with pernicious anaemia, in ten days the blood improved in the same way as if half a pound of liver had been eaten daily. If the gastric juice from a patient with pernicious anaemia was mixed with the beef steak and fed to another patient with pernicious anaemia, no improvement in blood resulted. It was obvious therefore that a failure of the stomach to secrete some essential factor was the cause of pernicious anaemia. This factor is now known to be an enzyme or ferment which acts on the proteins of the food and produces a substance which is stored in the liver and carried to the bone marrow to regulate the normal production of red blood corpuscles.

After these remarks on the history of the medical knowledge of the disease something may be written on its symptoms. The onset of the disease is usually gradual, and the first symptom is weakness. The patient complains of being easily tired. He may often be of an energetic disposition and refuse to submit to his feeling of weakness until his heart shows serious signs of overwork. Soreness of the tongue is another early symptom, especially in women. It occurs in about half of the cases.

The patient is often capable of working remarkably hard when in an advanced stage of the disease. This differentiates pernicious from other anaemias, and may be due to the slowness of the development which might allow the patient's body to become adapted to the condition.

The symptoms are often difficult to identify, because the disease occurs chiefly in and after middle age, and they may easily be confused with the effects of natural degeneration or senescence.

The existence of pernicious anaemia is not easily deduced from the patient's appearance. Sometimes he may have pallor, and sometimes his eyelids, feet and legs show dropsical swellings, and there are other characteristics, but none is general and striking.

The patient frequently complains of indigestion, and eats a badly balanced diet. Vomiting, diarrhoea and mild jaundice are not uncommon. Some authorities ascribe these not to the disease but to bad teeth or other sources of infection. Davidson and Gulland write that they have seen patients whose nausea was not affected by the removal of bad teeth. The pulse is usually rapid, over eighty, and goes up to one hundred and ten or twenty with excitement or slight exertion. The blood pressure is low.

Post mortems show that the kidneys are never quite normal, and albumen is not uncommonly found in the urine. Davidson and Gulland are of the opinion that the liver treatment may increase the amount of kidney disorders, as it increases the load of protein products of which the kidneys must dispose.

Menstruation nearly always ceases when the disease is developed, but it may return with an improvement due to treatment or otherwise.

Among the nervous symptoms are those due to disturbances in the spinal cord. These are exhibited by tingling and numbness in the hands and feet, and other extremities. About 80 per cent of cases show mild symptoms of this sort, but only about 5 per cent show definite evidence of damage to the cord. Besides symptoms due to organic lesions patients show various mental symptoms. They are often unexpectedly placid and contented. Some are of a contrary behaviour, excitable and neurasthenic, usually towards the end of the disease and in association with a relapse. The insanity is usually mild and often is exhibited by delusions of persecution. Relatives may become suspect. Davidson and Gulland write that they recall a lawsuit involving £500,000, in which a patient probably suffering from pernicious anaemia took a dislike to his family and left his fortune to charities. The will was disputed, but a compromise was arranged before court proceedings would have become necessary.

About two-thirds of the patients have fever temperatures at some stages of the disease, but the degree of fever is not directly related with the severity of the case. A sketch of the remarkable change seen in the blood of patients suffering from pernicious anaemia will be more intelligible after a short description of composition of some of the features of normal human blood. Normal blood is five times as viscous as water (blood is thicker than water) and has a slightly higher specific gravity. When examined under a microscope it is seen to contain a variety of objects. The most prominent are pale yellow discs. When in bulk these confer on blood its red colour. They are the red blood corpuscles. Their breadth is rather less than four times their thickness. Normal human blood contains about five million per cubic millimetre. If blood is allowed to stand in a paraffin-lined vessel the corpuscles settle into a red mass below a clear pale yellow liquid. The liquid is named plasma. After the whole has stood for twenty-four hours the red mass becomes jellified. The red corpuscles and a colloidal substance named fibrinogen from the plasma form a jelly. The abstraction of fibrinogen from the plasma leaves a clear liquid named serum.

Besides red corpuscles the blood contains a number of other objects revealed by suitable microscopic techniques. There are normally about eight or ten thousand white corpuscles in each cubic millimetre of blood. The white corpuscles are divided into two groups named leucocytes and lymphocytes. The leucocytes consist of four sub-groups named neutrophiles, eosinophiles, basophiles and monocytes; the first three owing their names to the manner in which they are stained by certain dyes. The neutrophiles are phagocytic, or devouring. They eat bacteria, and are mobilized at the site of wounds, which they keep clean. They also make the blood coagulate more quickly. The exact nature of their rôle in disease is disputed. The rôles of eosinophiles, basophiles and monocytes are even more obscure, though an important function of the monocytes is undoubtedly phagocytic.

These four types of cells and the red corpuscles are all formed in the marrow of the bones. The lymphocytes are chiefly produced by the lymph glands and are supposed to defend the body by destroying toxins in the blood. The white cells normally exist in the proportions:

Neutrophiles	.	.	.	.	70	per cent
Eosinophiles	.	.	.	.	3	,, ,,
Basophiles	.	.	.	.	0·5	,, ,,
Monocytes	.	.	.	.	4	,, ,,
Lymphocytes	.	.	.	20–25	,, ,,

A departure from these proportions is often the indication of the presence of disease.

Besides these cells and the red corpuscles there are a relatively large number of very small cells named platelets, which have an important part in coagulating blood. Each cubic millimetre of blood contains about 350,000.

The red corpuscles are grown in the marrow of the bones. They obtain the red pigment haemoglobin mainly from old corpuscles caught and decomposed by the cells of the spleen. Some of the products of decomposition are transported to the marrow and re-synthesized into haemoglobin. The cells that grow into red corpuscles have almost but not quite lost their nuclei before they are freed. The traces of immaturity in new red corpuscles have considerable importance in the detection of pernicious anaemia. The mature red corpuscles in the blood are not perfect cells but rather the finished products of cells. About one-fifteenth to one-thirtieth of the red corpuscles are lost every day, so the average life of a red corpuscle is about fifteen to thirty days. There is evidence that corpuscles transfused into the blood from another person may be detected by agglutination tests 100 days later, so the limit of corpuscular life may be very much longer than the normal.

The free red corpuscles are of remarkable uniformity in size, shape and condition. When immature red cells, still having well-developed nuclei, appear in the blood stream, disease is to be suspected. Irregularity of shape and size is another sign of disorder. The immature red cells are launched into the blood circulation in an attempt to supply a deficiency due to defect or loss of blood. The state of the blood in a person suffering from pernicious anaemia is in impressive contrast with the state of the blood in a healthy person. Instead of containing 5 million red corpuscles per cubic millimetre it may contain only 1 million, or less. The fall from 5 million to 1 million is usually slow and takes about six months or one year. The patient's body tends to become adapted to working with one-fifth the normal quantity of red corpuscles, and hence of haemoglobin. As the haemoglobin carries the oxygen from the lungs to the tissues the body behaves in some ways as if it were suffering from a lack of oxygen, as in a mountain climb where the air and oxygen are rare. The patient's weakness and breathlessness has a certain similarity with that of the high mountain climber's. The slow onset of pernicious anaemia allows the body to become adapted to working with a relatively small number of red corpuscles. This characteristic of the disease serves markedly to differentiate it from other forms of anaemia. Patients suffering from secondary anaemia due to cancer, blood poisoning or any other agency are quite unable to perform their daily work when the number of red corpuscles has been reduced to 1,000,000 per cubic millimetre.

The red cells are abnormally irregular in size. On the average they are about 15 per cent broader but the co-efficient of variation in diameter is much more impressive. In healthy persons the figure varies from 5 to 7 per cent, while in those suffering from pernicious anaemia the figure is from 8 to 18 per cent.

The presence in the blood of numbers of abnormally large red corpuscles is an important feature. When the number of corpuscles is down to 1,000,000 per cubic millimetre immature red corpuscles are seen, discharged from the bone marrow before their nuclei have disappeared. As previously explained, this is a sign that the bone marrow is working under strain.

The number of white cells in the blood is also changed. The total number of white cells in a cubic millimetre fall from about 9,000 in health to 5,000 or less. As previously mentioned, in health 70 per cent of the white cells are neutrophytes and 25 per cent are lymphocytes. In pernicious anaemia these percentages may be changed to 40 and 60 respectively. The direction of this relative change shows the bone marrow is disordered, because the neutrophiles besides the red corpuscles are grown in the bone marrow, whereas the lymphocytes come from the lymphatic glands. Hence pernicious anaemia must have more to do with the bone marrow than with the lymphatic glands. The difference in the effects of pernicious anaemia on the two groups of white cells provides evidence that the disease is not due, for example, to a poison or bacterium in the blood. If it were, the destructive agent would be expected to attack every sort of cell with more or less indiscrimination.

The nuclei of the neutrophile cells also show some abnormalities of structure.

The blood platelets are reduced from 350,000 per cubic millimetre in much the same proportions as the number of red corpuscles. When the latter number 1,000,000 the platelets number about 70,000.

The disorders in the stomach and intestines are almost as impressive as those in the blood. Davidson has compared the bacteria found in the faeces of professional footballers and students in athletic training with those in the faeces of persons suffering from cancer of the stomach, and other diseases, and from pernicious anaemia. It appears that though the footballers were in strict physical training they received no advice on diet from their trainer, and were quite without fads and interest in special foods. All of them were heavy meat-eaters, and some ate three meat meals daily. The bacterial content of their faeces was very low. The bacterial content of the faeces from the pernicious anaemia patients is much greater. There are more Bacillus coli, more streptococci, and very much more Bacillus welchii than in the faeces from normal patients or even in those suffering from such diseases as cancer of the stomach and severe secondary anaemia. Careful qualitative examinations of the Bacillus coli, welchii and streptococci from pernicious anaemia patients have shown that there are no special types associated with the disease. There is no evidence that pernicious anaemia is due to any bacillus. The conditions in which the disease occurs merely allow the bacilli normally found in human faeces to multiply beyond the usual numbers.

Similar results are obtained from the bacteriological examination of the contents of the stomach and intestines. No peculiar organisms are found in association with the disease, but there is a large increase in the numbers of the usual inhabitants of the bowel, such as B. coli, welchii and streptococci. Davidson writes that the large increase in the numbers of bacteria in the small intestine is often an important sign of the disease.

The large increase in the number of bacteria present in the intestines, and the presence of bacteria in the stomach, which is free from them in health, is due to the absence of free hydrochloric acid from the gastric juices. Bacteria cannot live in the strongly acidic juices secreted into a normal stomach. The pernicious anaemia patient swallows bacteria with his food, and from sores in his mouth, nose or throat. His stomach does not contain the agent that normally destroys the swallowed bacilli. There is also another source of infection. The muscles controlling the exit from the stomach become flabby, so that food is discharged from the stomach in half the normal time of one hour, and sometimes regurgitates back again. The regurgitation brings bacilli from the intestines into the stomach.

The absence of free hydrochloric acid from the stomach is one of the most certain signs of pernicious anaemia, but it is not a result of this disease. Patients that have become through liver treatment healthy enough to resume normal activities do not have free hydrochloric acid in their stomachs. Their stomachs and intestines still contain abnormal numbers of bacilli, in spite of which their health does not suffer severely.

Persons whose gastric juices do not contain free hydrochloric acid are probably liable to develop pernicious anaemia. An examination of the gastric juices of a hundred medical students showed in four cases that free acid was not secreted into the stomach. The connection between absence of acid and pernicious anaemia is obscure, but persons without acid should be watched as liable to develop the disease.

The administration of doses of hydrochloric acid has been recommended as a disinfectant of the stomach and a reducer of the excessive numbers of bacteria. It is not effective for this purpose as it is neutralized soon after entry into the stomach. The absence of free hydrochloric acid in the stomach may be due to neutralization by excessively alkaline mucous secretions. Though doses of hydrochloric acid do not achieve the acidification of the stomach, they may be effective for other ends such as the checking of diarrhoea.

There are other important digestive and assimilative disorders associated with the disease. During a relapse the patient usually loses more nitrogen than he absorbs. In a healthy adult man or animal the nitrogen content of the excreta is equal to the nitrogen content of the food. The body is unable to assimilate nitrogen from the air. It can use only nitrogen that has already been combined with other substances. No nitrogen is taken by the body from the air inspired into the lungs. Likewise no nitrogen is discharged from the body by the expired breath. The nitrogen passed through the lungs in breathing is inert, as in the cylinder of an internal combustion engine. The exhaust gases from an automobile engine contain just as much nitrogen as the air drawn into the engine through the inlet valves. The body must obtain the nitrogen required for building protein or fleshy materials from the food, as it cannot utilize the nitrogen in the air. As the nitrogen content of the expired breath is equal to that of the inspired breath, nitrogen obtained from food cannot leave the body in expired breath. It must be contained in substances dissolved in the urine, sweat and faeces. In health there is a steady equilibrium between the quantity of nitrogen absorbed from food and the quantity ejected in the excreta, except during growth, or when muscle is being laid on by physical training or continuous exercise. In convalescence after fevers and ill-health the quantity of nitrogen excreted is less than the quantity taken in food, as wastage is being repaired. In disease the patient may lose more nitrogen than he is absorbing, owing to the wasting of his tissues. This occurs in pernicious anaemia.

In 1917 Barker and Sprunt, and in 1918 Mosenthal, found that if patients were given forced feedings of diets rich in protein, i.e., fleshy materials, the nitrogen balance could be restored for a time, and the body might even begin to store nitrogen for the repair of wastage. The condition of the patient improves parallel with the improvement of the nitrogen balance, as shown, for example, by the increase in the number of red and other corpuscles in the blood.

This sketch of some of the aspects of pernicious anaemia may have helped to show the profound complication of the condition characteristic of the disease. There is an extraordinary change in the state of the blood. There are remarkable changes in the stomach and intestines, and there are notable changes in the nervous system. The investigator is always under a powerful compulsion to discover the explanation of things in the most accessible facts, whereas the true explanation may rest in other un-known facts. This tendency is responsible for the attempt to cure disease by treating the symptoms instead of discovering and treating the agent that has produced the symptoms. It leads to circular theorizing, in which one symptom is postulated as the primary agent and the others as derived from it.

Pernicious anaemia was recognized just over 100 years ago, and became a familiar clinical entity about sixty years ago. During the latter period it has provided a remarkable illustration of fallacious circular theorizing. The discovery of a satisfactory cure for the disease was not inspired by the researches which continued within the circles of the prominent symptoms. It was inspired by a scientific movement in progress without the circle of direct research on the problems of pernicious anaemia. The intellectual impulse towards the direction of a satisfactory treatment came from the modern researches on diet, of which vitamin physiology and chemistry is a famous branch. The creation of the modern refined conceptions of the influence of diet was chiefly due to Sir Frederick Hopkins. He had become an accomplished analytical chemist before he studied medicine, and did not start his hospital training until he was about thirty years of age. He entered the wards with a mind mature and naturally subtle and with an excellent scientific training. At that time tea-shops had been introduced in London. They were founded partly in response to the invention of the typewriter, as the introduction of this machine drew large numbers of girls to work in the city. The old public-houses were not suitable eating places for the girls, who wanted light and cheap meals in a less raw social environment. A house surgeon in Hopkins' hospital in the city of London had noticed how many of them came for treatment, and their low resistance to disease.

Though they did not complain of under-nourishment and appeared to have satisfied their appetites with the usual tea and buns, they seemed to have weak constitutions. The surgeon was convinced from experience that lemon juice strengthened them. Hopkins resolved that he would examine the action of the juice with scientific precision, when he had the opportunity. Twenty years later he was able to give the first completely satisfactory demonstration of the existence of vitamins. Before the modern recognition of the importance of the refined aspects of diet, the dietary treatment of patients suffering from pernicious anaemia was not regarded as of primary importance. The diet was conceived as a support to the body in its resistance to the disease; if the body was well-nourished it would withstand the disease more successfully. The problem lay in the provision of a diet digestible by a disordered stomach and intestines. Some investigators tried to discover diets that would remove prominent symptoms. As the quantity of red corpuscles in the blood is greatly reduced, the quantity of iron in the circulation is greatly reduced, as iron is an important constituent of haemoglobin, the red colouring matter in the corpuscles. Hence they recommended medicines or diets containing iron. This theory is fallacious, as pernicious anaemia is not due to a shortage of iron. One of the notable features at the post mortem of persons that have died from the disease is the excessive quantity of iron found in the liver.

Other remedies based on the treatment of symptoms were arsenic preparations and blood transfusion. Arsenic stimulates the activity of the bone-marrow so that more red corpuscles are produced and discharged into the blood stream. The condition of many patients was at first improved by arsenic, but after a time it relapsed and the efficacy of the drug became less and less after each application.

Transfusion of blood was first employed in the treatment of pernicious anaemia in 1880. When done with modern technique it is at first effective, but necessarily has no curative value, as the disease is due to defects in the mechanism of blood formation. When the cells of the new blood are worn out they are not adequately replaced. Hence transfusion has a temporary effect. But it may be of great assistance in the treatment of patients in whom the disease is advanced. They may have become too weak to digest the liver diet. A transfusion of blood may revive them for a week or so, and give their digestive systems strength to assimilate the large quantities of liver or liver extract necessary for permanent effective treatment.

Many early students of the disease had noticed the temporary value of changes in diet and air. The most remarkable of the early dietetic observations was that of Fraser in 1894. He had treated a patient with iron and arsenic without success, so he tried feeding him with raw bone-marrow from oxen and calves. The patient's condition improved rapidly. The number of red corpuscles increased from 1 to 4 millions per cubic millimetre, with parallel increases in the numbers of other blood cells. The effects of the marrow diet in this single example may have been illusory and the improvement merely an example of the inexplicable spontaneous recovery that sometimes occurs with pernicious anaemia, but Fraser was near a great discovery. If he had made his observation twenty years later in the atmosphere of vitamin dietary research he would have been freer from the misconceptions of symptom treatment. He was aiming at the treatment by diet of the bone marrow symptoms of pernicious anaemia. If he had been able to conceive the problem in the terms of the modern refined and quantitative conceptions of dietary research he might have studied the diet in much greater quantitative detail, and directed his attention away from a too exclusive study of symptoms. As in so many instances of medical discovery, the first satisfactory treatment of pernicious anaemia was discovered through experimental research on animals, and not from experimental treatment of patients. The human patient is often an inconvenient subject for experiment. He is human, and cannot be treated with severity, so the limits of any process of treatment cannot easily be found by trial. His illness makes his general physiology abnormal, so the experimenter often finds the interpretation of the effects of a treatment difficult. The doctor has a natural impulse to be more interested in curing the patient than in advancing scientific knowledge. The patient has usually engaged the doctor on that understanding. Strictly scientific methods of experimental research are applied much more easily to animals and even to healthy human beings than to sick patients. In the long run they are more effective than purely clinical research in leading to important new medical knowledge. The history of the treatment of pernicious anaemia provides an excellent illustration. Fifty years of clinical research provided a wide knowledge, indeed a bewildering knowledge, of the symptoms of pernicious anaemia, but the clues to an effective treatment came from general physiological research and not from special studies of the disease. An interesting summary of the history of dietetic research bearing on pernicious anaemia has been written by J. G. McCrie. He writes that in experiments done under the auspices of the Carnegie Institute of Washington in 1919 the effects of pronounced restriction of diet on twenty-four healthy persons were studied. The blood of most of the subjects showed a slight secondary anaemia due to the starvation.

The most fundamental researches were done by Whipple and Robscheit. In 1920 they published the results of some thorough investigations of the effect of diet on dogs in which anaemia had been produced experimentally. They measured the volume of blood in each dog and then produced a uniform degree of anaemia by removing on each of two successive days one quarter of the volume, so that each dog lost one-half of its normal volume of blood. The time taken for the blood to return to normal volume and condition and the effects of a variety of diets on the length of the time of recovery was noted. They found dogs fed on an ordinary diet of mixed table scraps recovered to the normal condition in four to seven weeks. Dogs fed on a liberal diet of meat and beef heart recovered in three to four weeks, and dogs fed on cooked liver recovered in two to four weeks. They found extracts of liver made with water had a slight but distinct effect in regenerating blood, while commercial meat extract had no effect. The food materials that helped to regenerate blood were effective alone or when mixed with other foods, and they would even exert their influence when given after long periods of diet unsuitable to blood regeneration.

In 1925 Whipple and Robscheit-Robbins described an extension of these researches. They kept dogs in a constant state of severe secondary anaemia by frequent bleeding. They found the dogs in this condition recovered most rapidly when fed on beef liver, less rapidly on beef heart, and still less rapidly on beef muscle. They concluded liver feeding is the most effective agent of blood regeneration in severe anaemias of the type in their experiments. They remarked that its beneficial action was invariable and independent of the other constituents of the diet, or the length of the period during which the dog had suffered from the severe anaemia.

Their emphatic conclusions gave a powerful stimulus to the clinical investigation of anaemia. The power of the method of experiment on animals to advance medical discovery is illustrated magnificently by the work of Whipple and Robscheit. As a result of a long series of exact experiments on animals in a uniform condition they demonstrated with convincing certainty that a diet of beef liver is a very effective regenerator of the blood. Human convention precludes the performance of series of similar experiments on human subjects, so the discovery of the liver treatment from direct research on human beings would have been extremely improbable. When the discovery has been made nothing seems simpler. But who would have expected that a diet containing half a pound of liver daily would be an effective treatment for persons suffering from the many disorders present in pernicious anaemia? Could their weakened digestive systems have been expected to assimilate daily such large quantities of liver, and would not the experiment of trying such an unpalatable diet have been regarded as cruel, if the experimenter had no evidence that it might be effective? Animal experimentation so often provides the definite knowledge that gives the clinician confidence to try new treatments on human beings, the effectiveness of which would appear to be very improbable without hint.

Under the influence chiefly of the results of Whipple and Robscheit's experiments Minot and Murphy fed a number of pernicious anaemia patients on a daily diet containing from a quarter to half a pound of cooked beef liver, quarter of a pound of red meat, not less than five or six ounces of vegetables, quarter of a pound of milk, and dry and crusty bread. Little sugary food was included. Minot and Murphy had noticed that pernicious anaemia occurred less frequently in populations that eat less than the usual proportion of starchy food. They believed the disease might depend on some nutritional factor, and commented on the value of strawberries in the diet of persons suffering from sprue, a tropical disease with symptoms very similar and sometimes almost indistinguishable from those of pernicious anaemia. Of the first forty-five patients given the diet, four died. These were so ill that improvement could not reasonably be expected. Within two weeks the rest were eating the diet with ravenous appetite. Whereas normal persons might have been nauseated with so much liver and beef, the pernicious anaemia patients continually ate the complete rations with zest. At the end of the first week the blood showed improvement, and at the end of four to six months a third of the patients had the normal number of red corpuscles, and no patient had less than 70 per cent of the normal.

Further experience showed that the liver was the important part of the diet.

In 1927 Cohn showed an extract potent for pernicious anaemia could be made from liver. The extract was soluble in water, but insoluble in ether; it contained virtually no iron or protein and could be percipitated by alcohol. This discovery showed the active factor in liver must be of a relatively simple character, and probably in the future would be isolated and synthesized in the laboratory. Cohn showed later that the factor could be precipitated from liver with phosphotungstic acid, and he obtained a substance 0·6 grm. per day of which was sufficient to restore the patient. Also, it enabled patients that had difficulty in digesting heavy liver diets to take the active liver substance in a concentrated form. A small quantity of the extract was as potent as the large volumes of liver. Within the last two years liver extracts for injection have been used. Since this method is thirty times more effective than liver or liver extract taken by the mouth, it is particularly valuable during the acute stage when the patient is very ill. A new preparation made from the stomach of hogs has recently been introduced. It is said to be more effective and cheaper than liver or liver extract.

After further experience Minot and Murphy stated that if a patient did not respond to liver treatment within six weeks a mistake had been probably made in the diagnosis of his disease.

There is little evidence that the liver treatment when given under proper supervision produces fresh secondary disorders. Davidson and Gulland write that the public should be warned against an indiscriminate eating of liver. The liver treatment is of full potency only for pernicious anaemia. It is much less effective for other sorts of anaemia. The existence of an effective treatment for a disease often produces a tendency in doctors to mistake the symptoms of other diseases as symptoms of the easily treatable disease. Doctors now occasionally diagnose a severe anaemia as pernicious anaemia and recommend the liver treatment when the anaemia is of a secondary character due, for example, to cancer of the stomach.

Davidson and Gulland have also commented that poor people find the cost of the liver treatment a severe tax on their means. The discovery of the value of a liver diet for patients suffering from pernicious anaemia, which is one of the most brilliant achievements of scientific medicine, has had the ironical effect of making liver more difficult for poor patients to obtain.

The price of liver has been greatly increased by the demand created by the new treatment. Pernicious anaemia patients should be assured by law of an adequate supply of liver at a reasonable price.

CHAPTER X

ARTIFICIAL RADIOACTIVITY

THE late Mme. Curie lived just long enough to see the brilliant discovery of artificial radio-activity by her daughter and son-in-law, Irene and F. Curie-Joliot.

Those interested in the conditions of the progress of science will do well to study the history of her career. The famous discovery of radium was made in collaboration with her husband in 1898. The brilliant discovery of artificial radioactivity was made by her daughter and son-in-law in 1934—thirty-six years later. Compare these dates with those of Rutherford's career. He saw how radioactivity could be turned to the analysis of the structure of atoms in 1903–5, and announced his determination to make the research. In 1932 his colleagues Cockcroft and Walton showed how atoms could be disintegrated by machinery. This parallel suggests that a school and tradition of supreme research requires a generation for growth. It is difficult in less than a generation to accumulate resources and to perfect a multifarious technique. When a school of research has grown during thirty years it can solve an astonishing number of problems, because any new fact can be investigated in a dozen different ways by a group of experienced and accomplished scientists. While the school is being founded it is led by a brilliant investigator with moderate resources and a small number of students.

The leader will make great discoveries, but his colleagues will as a rule be helpers rather than original discoverers. The subject will be in an early stage of development and present few aspects sufficiently developed to be susceptible of swift solution. After one or two decades various aspects will become more clearly significant and the technique for investigating them will have grown. After a generation new problems will be solved soon after they appear, and the school will in a short period make an extraordinary number of important discoveries. The school possesses a mental attitude that enables it to recognize phenomena significant to the theme of its researches. The recent history of the schools of radioactive research at Paris and of atomic physics at Cambridge illustrate these tendencies. Paris is pre-eminent in those aspects of recent radioactive research that resemble the aspects of radioactive research explored by Mme. Curie thirty years ago. The tradition she has cultivated has given her school a special facility in those types of radioactive research which she had herself so much developed. The Rutherford school now established at Cambridge is pre-eminent in the study of atoms and atomic structure, and has a special facility for recognizing the atomic aspect of new knowledge.

The discovery of the neutron has been described in Chapter I. Chadwick's courage in postulating the existence of the neutron started a new period in physical research. In the intellectual excitement following his discovery, Cockcroft and Walton were inspired to attempt the disintegration of atoms by machinery, and succeeded. Investigators throughout the world began to look for new particles. Anderson published some evidence for the existence in connection with cosmic rays of the positive electron or positron, which was confirmed by Blackett and Occhialini early in 1933. Chadwick, Blackett, and Occhialini discovered laboratory methods of producing positrons. They had noticed the Curie-Joliots' observation that negative electrons sometimes seemed to be moving backwards in a Wilson chamber traversed by the mixture of neutrons and wave-radiations from bombarded beryllium. They considered these particles might be positive electrons moving forwards from atoms struck by the neutrons or waves, and showed experimentally that positrons could be produced by bombarding pieces of lead with the mixed radiation from beryllium.

The production of positrons by bombarding atoms with neutrons or wave-radiations was of remarkable interest in connection with Dirac's theory of positrons. He conceives the positron as a hole in the universe marked by the absence of a negative electron. When a negative electron drops into this hole two units of wave-radiation are emitted. Conversely, wave-radiation is capable of being converted back into pairs of positive and negative electrons. The Curie-Joliots named this process 'materialization', as particles were produced out of waves. They found, also, that positrons might be produced by another sort of process, as a by-product of the artificial disintegration of atoms of aluminium. They found that atoms of aluminium and other light elements emitted positrons when bombarded with the swift helium atoms ejected from polonium. While investigating this phenomenon they discovered a very remarkable fact. The emissions of positrons from aluminium does not cease immediately after the removal of the polonium which provides the bombarding helium atoms. The bombarding helium atoms do not simply disintegrate the aluminium atoms; they transmute them into a state in which they can emit positrons after an interval of time. The rate of emission of positrons after the polonium has been removed decreases in the same way as the rate of emission of particles from natural radioactive substances. This means that the bombardment converts the aluminium foil into a radioactive condition. This was the grand discovery of the possibility of making artificial radioactive substances. Like the discovery of radium, it has immense practical besides theoretical interest. This aspect will be discussed presently. The Curie-Joliots immediately discovered that boron and magnesium could also be converted into a radioactive condition by bombardment with helium atoms obtained from polonium. They found that the number of positrons produced from bombarded boron was reduced by 50 per cent after 14 minutes, the corresponding figures for magnesium and aluminium being 2 minutes 30 seconds and 3 minutes 15 seconds respectively.

Besides providing the first examples of artificial radioactivity these experiments proved the existence of a new sort of radioactive radiation. Hitherto, three sorts of radiations from radioactive substances had been recognized: helium particles, negative electrons, and waves. The Curie-Joliots showed that positive electrons could also constitute a radiation from a radioactive substance. A consideration of the structure of the boron atom suggests that it might produce an unstable form of nitrogen when bombarded by a swift helium atom.

Then the Curie tradition exerted its influence. Like Mme. Curie thirty-six years ago, the Curie-Joliots immediately attempted to separate the new radioactive substances by chemical methods. They bombarded the substance boron nitride, a compound of boron and nitrogen. When boron nitride is treated with caustic soda gaseous ammonia is produced. They found that the radioactive material formed in the boron during the bombardment is carried away with the ammonia. Hence the radioactive material was almost certainly a form of nitrogen, because nitrogen is one of the two constituents of ammonia. The other constituent is hydrogen, and it was very improbable that the hydrogen had been made radioactive. They found that the active substance in bombarded aluminium could be separated by solution in hydrochloric acid. It was carried away with the hydrogen released from the acid, and appeared to form phosphine. It could be precipitated by zirconium phosphate. These experiments indicated that it was a radioactive form of phosphorus. The magnesium appeared to give a radioactive form of silicon. The Curie-Joliots named the three new substances radio-nitrogen, radio-silicon, and radio-phosphorus.

The manufacture of radioactive substances by bombardment with helium atoms suggested that bombardments with other sorts of particles might be effective—in particular, particles accelerated by electrical machinery. J. D. Cockcroft and E. T. S. Walton had made the famous experiments on the disintegration of atoms by machinery. It was not surprising that these masters should inquire whether radioactive substances could be made with their machines. With the collaboration of C. W. Gilbert they discovered that ordinary carbon when bombarded by hydrogen atoms, or protons, was converted into a radioactive condition. The number of positrons emitted from the bombarded carbon decreased by 50 per cent after the lapse of 10 minutes. They concluded that the carbon atoms were transmuted into atoms of radio-nitrogen by the incorporation of the bombarding protons.   Within two months of the discovery of artificial radio-activity a radioactive substance had been manufactured by electrical machinery out of two of the commonest substances, carbon and hydrogen.   It is impossible to exaggerate the wonder of this achievement.   Since the discovery of the neutron in 1932, the initiative has passed from the theorists to the experimenters.   During the last two years the experimenters have discovered the neutron, the positron, the artificial disintegration of atoms by machinery, artificial radioactivity, and the manufacture of radioactive substances out of common materials by machinery.   Other astonishing experimental discoveries will be described presently.

Is it surprising that the excitements produced by the implications of the principle of uncertainty have declined?

For six years after 1927, when Heisenberg formulated the principle, there was much discussion of its implications, and how they might indicate the limits of scientific inquiry.   A belief arose that the principle implied that scientific inquiry had unexpectedly narrow limits.   Many asserted that the scope of scientific progress was lessened, and that the aspirations of scientists must be curtailed.   Undoubtedly the reports of these discussions that reached the public persuaded many to believe that the methods of scientific discovery were less important and effective than had previously been supposed.   The prestige of the scientific method had been reduced.   Everywhere discussions on the limitations of science were heard. These discussions have become a little lame after the almost unparalleled riot of experimental discovery during the last two years.

The first chemical separation of new artificial radioactive substances performed in England was done by O. R. Frisch. He found that sodium and phosphorus could be converted into a radioactive condition by bombardment with swift helium atoms.

The radiations emitted by the artificial radioactive substances prepared by the Curie-Joliots consisted of positrons. Alichanow, Alichanian, and Dzelepow, of Leningrad (whose laboratory I have seen, and who kindly explained to me their ingenious apparatus), have found that an artificial radioactivity which emits ordinary negative electrons may also be obtained. If magnesium is bombarded with swift helium atoms it is converted into a radioactive condition in which positrons are emitted, as the Curie-Joliots found. But Alichanow has shown that negative electrons also are emitted by the activated magnesium, the number being about four times the number of emitted positive electrons.

In their original paper the Curie-Joliots had suggested that radioactivity might be induced in carbon by bombardment with atoms of heavy hydrogen. This suggestion was confirmed by Henderson, Livingston, and Lawrence, of Berkeley, California. They bombarded carbon and many other substances with atoms of heavy hydrogen accelerated to an energy of 3,000,000 volts. They found evidence of the production of radioactivity by this treatment in carbon, calcium fluoride, calcium chloride, lithium carbonate, ammonium nitrate, aluminium nitrate and magnesium. Two radioactive substances appeared to be produced in beryllium, one with a half-life of 9 minutes and the other of 3 minutes. Crane and Lauritsen, of Pasadena, obtained radioactive gases from carbon and boric oxide by bombarding them with heavy hydrogen. These gases are probably radio-nitrogen.

Atoms of helium, hydrogen, and heavy hydrogen had proved to be efficient projectiles for producing artificial radioactivity in various substances, but only in light elements such as carbon, aluminium, and boron. They had not transmuted atoms of the heavier elements, such as iron. This was to be expected, as the bombarding particles carry a positive charge of electricity. They would not be able to approach the nuclei of the heavy atoms because they would be repelled by the large positive electrical charges on the nuclei of the atoms of the heavy elements. The limitation of the hydrogen, heavy hydrogen, and helium particles, in virtue of their electric charges, does not apply to a neutron. As it has no charge it should have no difficulty in penetrating the nuclei of all atoms, and perhaps of producing rearrangements leading to the formation of new radioactive nuclei. Professor Enrico Fermi, of Rome, began experiments to see whether neutrons might be efficient producers of artificial radioactivity in other substances. The results of his experiments are astonishing. Scientists were peculiarly interested by them, because Fermi has not hitherto been known as an experimenter. He is one of the world's leading theoretical physicists, and is thirty-three years of age.

Fermi prepared a convenient source of neutrons by sealing a mixture of radium emanation and beryllium powder in a glass tube. The helium particles emitted by the spontaneously disintegrating radium emanation eject neutrons from the beryllium atoms. Tubes emitting 1,000,000 neutrons per second may be made in this way. With the collaboration of D'Agostino, Amaldi, and Segrè he obtained a remarkable series of results. Phosphorus became strongly radioactive under the neutron bombardment. The electrons emitted by the disintegrating substance produced in the phosphorus could be photographed. Chemical separation showed that the substance was probably radio-silicon. Iron provided a radioactive substance with a half-life of about 3 hours. Chemical treatment showed that it was probably radioactive manganese. Aluminium gave a substance with a half-life of 12 minutes, and the disintegrations could be photographed. Silicon gave a substance with a half-life of about 3 minutes, and the disintegration electrons were photographed.

Chlorine gave a substance with a half-life much longer than that of any other substance produced in the series of experiments. Vanadium gave a substance with a half-life of 5 minutes. For arsenic the half-life is 2 days. A strong effect was obtained with silver, the half-life being about 2 minutes. Tellurium gave a period of about 1 hour. Iodine gave an intense effect with a half-life of 30 minutes. Chromium gave an intense effect with a half-life of 6 minutes, and the disintegration electrons were photographed. Barium gave a half-life of 2 minutes, and fluorine of 10 seconds. Effects were obtained also with sodium, magnesium, titanium, zirconium, zinc, strontium, antimony, selenium, and bromine.

All of these results were mentioned in a short letter to *Nature*. The vast extension of new facts made the scientific imagination reel. They were followed by even more results of an equally remarkable kind. Fermi and his colleagues found that at least 47 elements could be converted into radioactive substances by neutron bombardment. They investigated the sign of the electric charge on the electrons emitted by the disintegrating substances and found it was always negative. All of the new radioactive atoms produced by neutrons emitted negative electrons. This was in contrast with the new radioactive atoms produced by helium bombardments, which emitted positive electrons. They found that the atomic number of the substances produced in aluminium, chlorine, and cobalt was two less than that of the parent atoms. In four cases—phosphorus, sulphur, iron and zinc—it was one less. In the cases of bromine and iodine the active element is an isotope of the bombarded element, i.e., the atoms of bromine and iodine are converted from inactive into active atoms, but their chemical properties remain unchanged.

With the assistance of Rasetti and D'Agostino, Fermi investigated the interesting problem of the effect of neutron bombardment on the atoms of uranium and thorium. These are the very heavy natural radioactive substances. The experimenters thought that a series of new disintegrations might be started by a neutron bombardment in these naturally unstable atoms. Their experiments showed that this was correct. Uranium and thorium both showed a large increase in activity after bombardment. Two new radioactive substances appeared to be produced in the thorium, and five in uranium. The half-lives of three of the uranium substances are 10 seconds, 40 seconds, and 10 minutes. The half-lives of the remaining pair are between 40 minutes and 1 day. The 13-minute substance can be separated by chemical treatment. The uranium preparation is treated with nitric acid, and a small quantity of a manganese salt is added. The manganese is precipitated in manganese dioxide by boiling the solution with sodium chlorate. A large part of the active substance remains in the manganese dioxide. This chemical treatment shows that it cannot be a form of thorium or palladium, or bismuth or lead or ekacæsium or emanation. The atomic number of uranium is 92, and those of the other elements mentioned are 90, 91, 88, 89, 83, 82, 87, and 86. If the 13-minute substance is not one of these, what can it be? Perhaps it is a uranium atom that has incorporated part of a neutron, and increased its mass from 92 to 93 units. Such a substance should have chemical properties resembling those of manganese, according to the periodic law of the chemical properties of the elements. As the 13-minute substance seems to follow manganese through chemical reactions, it may be element 93.

The number of elementary atoms known to chemistry was 92, the lightest being hydrogen and the heaviest uranium. Some theorists had offered arguments to show that not more than 118 could be possible. Later the maximum number was reduced to 96, and recently two physicists argued that the maximum number was 92. The apparent experimental discovery of element 93 is rather awkward for them! Fermi thinks that element 94 or 95 may also be produced in the neutron bombardments. The heaviest atom found in the earth is that of uranium. Heavier atoms may have existed at some time in the past, but they disintegrated aeons ago. Jeans has suggested that radioactive atoms of masses 94 or more might exist in the stars and provide some of the stellar energy by disintegration. If Fermi's substance is element 93, he will be the first man to have made an element. His predecessors have only discovered chemical elements or prepared radioactive varieties of known elements. The manufacture of elements provides a new attitude in chemical science. Hitherto chemists investigated the properties of atoms provided by Nature. Now they have the prospect of investigating the chemistry of an atom prepared by themselves. While the chemistry of No. 93 will resemble that of certain other elements, according to the periodic law, it will in detail be unique. Preparing a new atom implies the invention of a new chapter of chemistry. The problems of the chemist were previously set him by Nature. He is no longer entirely dependent on Nature for his problems. He can invent them for himself, by manufacturing atoms not found on the earth, or observed in the heavens.

The disintegration of atoms by charged particles such as helium nuclei and protons, and neutral particles such as neutrons, has been followed by the accomplishment of disintegration by wave-radiations. J. Chadwick and M. Goldhaber have found that the diplon can be disintegrated by very energetic γ-rays. They filled an ionization chamber with heavy hydrogen and submitted it to γ-rays emitted from a source of radio-thorium. Electrical recorders showed that protons were produced in the chamber by the γ-rays, and must have arisen from the disintegration of diplons. L. Szilard and T. A. Chalmers have made an important addition to the method of disintegration by wave-radiation. They have found that when beryllium is submitted to γ-rays from radium it emits a radiation which can induce radioactivity in iodine. They conclude that the γ-rays from radium disintegrate the beryllium atoms and make them emit neutrons. The neutrons fly into the iodine atoms and transmute them into radioactive atoms, according to the Fermi effect. In the case of iodine, the atoms are transmuted into radio-iodine which may be separated by suitable chemical treatment. These experiments show that artificial radioactivity may be produced by the γ-rays from sealed radium containers. These are used in many hospitals for administering therapeutic treatment, and will therefore provide a convenient and common source for producing artificially radioactive substances. Szilard and Chalmers have explained that their method of employing wave-radiations for producing artificial radioactivity through the medium of neutrons should enable powerful X-ray tubes, providing X-rays as energetic as γ-rays, to be used as sources for the production of artificial radioactivity. The X-rays would be used for ejecting neutrons from beryllium, and the neutrons would then be used for producing Fermi transmutations of inactive into radioactive atoms.

What are the implications of the indescribable burst of physical discovery during the last two years? Until about six months ago 92, or perhaps only 91, chemical sorts of atoms were known. Most of these were stable atoms, such as those of oxygen, or iron or gold. The stable atom was regarded as the typical atom. Unstable atoms, such as those of radium, were known, but they were regarded as exceptional. During the last six months the number of known sorts of physical atoms has been almost doubled; some evidence has been found for at least 60 new isotopes. Is it not possible that the old idea that normal atoms are stable, while unstable atoms are abnormal, may have to be reversed? In the future hundreds of sorts of unstable atoms may be discovered, and the 100 stable old atoms and their isotopes, such as those of oxygen, iron and gold, will be regarded as abnormal. Ordinary stable atoms will be regarded as singularities and remainders in a vastly more multifarious possible world of unstable atoms. At some period when the universe was very young it contained more than 92 sorts of chemical atoms, and hundreds of unstable varieties of these atoms. The interactions of matter and radiation at that time were the exception and unstable atoms were the rule. That universe would to an observer have appeared extremely dynamic, and our own universe of slow chemical reactions between stable atoms would in comparison appear almost deadly static.

The practical value of the manufacture of radioactive substances may become great. The natural radioactive substances, such as radium, are rare and expensive. Radium costs about £300,000 per ounce. The rarity and commercial value of radium increase the difficulty of investigating and applying its medical properties. The new artificial radioactive substances may be more effective and more convenient than those so laboriously ground out of the rocks. The improvement in the technique of their manufacture will in future probably make them available in large quantities at reasonable prices. They may have interesting biological applications. Radio-nitrogen might be used to trace chemical and biological reactions by substituting it for ordinary nitrogen in protein and other important substances. As it leaves no dangerous radioactive deposit, it could be safely introduced by its chemical affinities into the living body, and reach regions at present inaccessible to radioactive treatment. While many persons will be impressed by visions of future radioactive manufactories, and the applications of the products in medicine and in goods for human convenience, others will be impressed by the sudden multiplication of the aspects of the material universe. Almost every month the theorist finds new fundamental facts rising before him in profusion. What new conceptions must he devise to comprehend the new facts, and in what perspective will the problem of the nature of the universe and existence be placed? Shall we in a few years' time reach a conception of the evolution of the material universe profoundly different from the one which we have at present? Theories of the internal constitution of the stars must be deeply modified by the discoveries of the new science—the chemistry of the nuclei of atoms. Atoms of lithium can be disintegrated by protons travelling under the electrical pressure of 30,000 volts. This is quite a trifling voltage. The electrical transmission lines from the Boulder Dam power station being erected in the canyon of the Colorado river in America will operate at 287,000 volts, i.e., at a tension almost ten times that sufficient to disintegrate lithium. If the voltages of mundane human power stations are commonly to be so much greater than the voltages necessary to disintegrate certain atoms, what must be happening inside stars, where the physical conditions are so extremely violent? Atoms must be disintegrated and the parts recombined in a turmoil of reactions. The complication of the details of these reactions will require vast extensions of astrophysical theory for its description.

The common use of voltage so very much higher than those necessary to make some atomic disintegrations inevitably suggests that immense developments in the human control of Nature will arise before many years have passed. But it is also certainly true that scientists cannot at present see how these developments can be made. Scientists are enjoying the exhilaration of swift advance, and approach the problems of Nature with more confidence than ever. The pessimism derived some years ago from inadequate discussions of the implications of recent theoretical discoveries is shown to be unnecessary. Scientists are acquiring new knowledge at an astonishing pace. But they are anxious that civilization shall not collapse and that humanity shall not fail to benefit from the new knowledge. Everyone who appreciates the achievements of contemporary science must be anxious that civilization should be ordered in a way that these achievements shall not be wasted.

INDEX

313

Milton Keynes UK
Ingram Content Group UK Ltd.
UKHW031142141024
449569UK00024B/1137